Advances in Japanese Business and Economics

Volume 30

Advances in Japanese Business and Economics (AJBE) showcases the work of Japanese and non-Japanese scholars researching the Japanese economy and Japanese businesses. Published in English, the series highlights for a global readership the unique perspectives of Japan's most distinguished and emerging scholars of business and economics. It covers research of either theoretical or empirical nature, in both authored and edited volumes, regardless of the sub-discipline or geographical coverage, including, but not limited to, such topics as macroeconomics, microeconomics, industrial relations, innovation, regional development, entrepreneurship, international trade, globalization, financial markets, technology management, and business strategy. At the same time, as a series of volumes written by Japanese and non-Japanese scholars studying Japan, it includes research on the issues of the Japanese economy, industry, management practice, and policy, such as the economic policies and business innovations before and after the Japanese "bubble" burst in the 1990s.

AJBE endeavors to overcome a historical deficit in the dissemination of Japanese economic theory, research methodology, and analysis. The volumes in the series contribute not only to a deeper understanding of Japanese business and economics but to revealing underlying universal principles.

Overseen by a panel of renowned scholars led by Editor-in-Chief Professor Ryuzo Sato, AJBE employs a single-blind review process in which the Editor-in-Chief, together with the Managing Editors and specialized scholars designated by the Editor-in-Chief or Managing Editors, rigorously reviews each proposal and manuscript to ensure that every submission is a valuable contribution to the global scholarly readership.

All books and chapters in AJBE are indexed in Scopus.

More information about this series at http://www.springer.com/series/11682

Kimiko Terai · Amihai Glazer ·
Naomi Miyazato

The Political Economy of Population Aging

Japan and the United States

Springer

Kimiko Terai
Faculty of Economics
Keio University
Tokyo, Japan

Amihai Glazer
Department of Economics
University of California
Irvine, CA, USA

Naomi Miyazato
College of Economics
Nihon University
Chiyoda, Tokyo, Japan

ISSN 2197-8859 ISSN 2197-8867 (electronic)
Advances in Japanese Business and Economics
ISBN 978-981-16-5535-7 ISBN 978-981-16-5536-4 (eBook)
https://doi.org/10.1007/978-981-16-5536-4

This Springer imprint is published by the registered company Springer Nature Singapore Pte Ltd.
The registered company address is: 152 Beach Road, #21-01/04 Gateway East, Singapore 189721, Singapore

Preface

Population aging is a major topic of research. It is also a key policy issue in developed countries. This book integrates the economics of aging and the insights of political economy. Some researchers are concerned about the changing structure of the labor force, a decline in saving, and their effects on economic growth. Others focus on the increased fiscal burden that will be ultimately imposed on younger generations to maintain health care, pension, and long-term care systems. This book addresses the salient question: what is the consequence of changed political powers across generations with regard to policy adoption and welfare?

Political-economic analyses using formal models describe how citizens' policy preferences are aggregated through the political process. These theoretical analyses show that the preferences of decisive voters determine polices. If voters' policy preferences depend on their age, a shift in the electorate's age structure changes the identity of pivotal actors, and therefore the policies adopted by government. We investigate such problems by conducting comparative studies using Japanese and US panel datasets.

Some characteristics create differences in policy preferences by age. Following the life-cycle consumption hypothesis, citizens at different life stages have different incomes, assets, and number of dependent relatives. Consequently, they differ in what they want from government. The book also discusses the discrepancy between the time period for which an existing policy generates positive effects, and the life expectancy of a citizen; the latter is mainly determined by his/her age. This generates an externality problem in a dynamic setting. Voters with short life expectancy would not take into consideration long-term benefits. In an aged society, the elderly are pivotal voters whose favored policies are likely to be adopted, possibly neglecting the effects of policy on the current young and on future generations.

As an illustration, consider investments in human and physical capital. Spending for these purposes may benefit younger citizens and future generations more than the elderly. With the increase in political power of elderly citizens and constraints on public resources, resources may be moved from those benefiting younger citizens, such as public investment programs and education, to policies that benefit the elderly, such as welfare. And taxes may be small on the elderly.

We conduct statistical analysis to investigate whether regions with a higher proportion of the elderly population adopt policies that generate short-term benefits, instead of policies that benefit younger generations for a long period. Based on the estimation results, Japan and the US are compared with regard to the political influences of elderly voters and the effects of differentiated institutional characteristics. In particular, the results reported in the book indicate that institutional conditions affect how the elderly may influence policy.

The fiscal system in Japan is more centralized than in the US. A decentralized fiscal system can also sort citizens according to their preferences. Residents can move to other regions that provide public goods or transfer programs that they prefer. Sorted residents can induce the regional government to implement policies that match their preferences more closely. For instance, younger citizens may be drawn toward regions having a younger population where public services such as education are supported. Thus, the aging population structure of the economy may not impair growth.

In contrast, lower residential mobility induced by the centralized fiscal system, as in Japan, could induce young residents to care about the physical capital of the region. Our estimation results support this conjecture. Japanese prefectures with a higher ratio of aged population invest less in physical capital.

Although we rely on statistical methods for comparative analysis, hypothesis testing and problem insights are derived from economic theory. Instead of presenting mathematical expressions, we highlight our research objective, perspectives on changing societies, and suggestions on institutional reforms, in a descriptive manner using diagrams wherever necessary.

The book provides economists and political scientists with guidelines for problems arising from an aging population. Moreover, each chapter explores the influence of increased political power of elderly voters on expenditure, regulation, and taxation by government, with other political and institutional factors controlled. Therefore, the book is a concise collection of empirical works that explore determinants of policies for all readers, including graduate students studying public economics, local public finance, and political economy. In addition, the book provides practical insights for policymakers.

We launched our joint research at the University of California, Irvine (UCI). Two authors of this book, Kimiko Terai and Naomi Miyazato, had the opportunity to stay at the UCI for two years as visiting researchers. Both got valuable insights and guidance from Amihai Glazer, UCI, who is another author of this book. After our sabbatical, we often had the opportunity to discuss the various problems in Japan and the US, which prompted us to collaborate on this project. We are grateful for the resources and ambience provided by UCI, which included delightful discussions with other professors during lunch hours and the hospitality of staff. We are also thankful for the comments and suggestions given by many researchers during seminars and academic meetings, including those of Professor Toshihiro Ihori, Professor Kyota Eguchi, Professor Mizuki Kawabata, Professor Shigeki

Kunieda, and Professor Michio Yuda. Their comments and suggestions helped us to significantly improve the manuscript. Last but not least, we are deeply indebted to Professor Ryuzo Sato, Editor in Chief of Advances in Japanese Business and Economics series, Springer, for giving us an opportunity to publish this book, and Ms Juno Kawakami, Springer, for giving us effectual editorial advice and for guiding us toward the publication.

Tokyo, Japan Kimiko Terai
Irvine, USA Amihai Glazer
Tokyo, Japan Naomi Miyazato

Contents

Chapter 1
Introduction—Population Aging and the Political Economy

1.1 Basic Idea

Many developed countries are facing problems associated with the rapid aging of their populations. If voters' policy preferences depend on their age and the government's policy adoption reflects public opinion, the change in demographic composition may affect governmental policies in a region. Therefore, to retain their political power, ruling parties often reallocate the budget in favor of policies preferred by the elderly, rather than younger generations. This book investigates such outcomes by focusing on whether regions with aged populations largely adopt policies that generate benefits in the short term at the expense of policies that can benefit younger generations in the long term. In other words, regions with aged populations may spend more on transfers and consumable public goods, rather than on the accumulation of physical and human capital.

One way to measure a region's extent of population aging is to calculate the percentage of the population aged 65 years and over. Figure 1.1 depicts the post–World War II values of this index for selected Organisation for Economic Co-operation and Development (OECD) countries. The figure indicates the following. First, many of the countries are experiencing population aging. Second, among these countries, Japan occupies a prominent position in both the extent and speed of population aging. Hence, a study of Japan may indicate what will happen later in other countries.

To examine why and how demographic structure affects governmental policy, we build on previous findings in the field of political economy. The median voter theorem (Black 1948) or the theory of electoral competition between two opportunistic political parties (Downs 1957) predicts that if voters have single-peaked preferences and if the policy space is unidimensional, the bliss point of median voters will be adopted through the political process. The main issues in electoral campaigns are not always unidimensional, which indicates that assuming a single issue dimension may be inappropriate. However, the idea that politicians pander to the electorate whose opinion is moderate seems relevant. In this case, if citizens' policy preferences vary throughout

K. Terai et al., *The Political Economy of Population Aging*, Advances in Japanese Business and Economics 30, https://doi.org/10.1007/978-981-16-5536-4_1

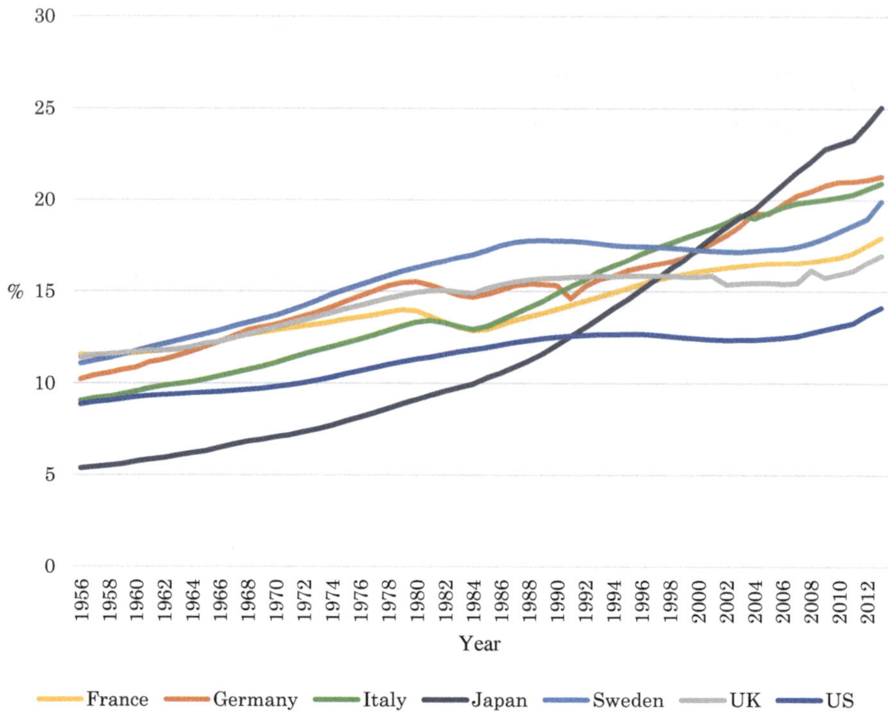

Fig. 1.1 Percentages of population aged 65 years and over in different countries. *Data* OECD, Elderly population (indicator). https://doi.org/10.1787/8d805ea1-en (Accessed on 30 May 2019)

their life cycles, the government is more likely to implement policies that benefit the elderly as median voters grow older.

In our analysis, we focus on the duration for which an implemented policy continues to generate positive effects. Since the life expectancy of the elderly is shorter than that of younger generations, the former generally prefers policies that generate benefits within short periods over policies that generate benefits for long periods. Based on this perspective, we test the following hypotheses.

- Aging economies spend little on education; investment in education is interpreted as an investment in human capital.
- Aging economies spend little on construction and long-lived infrastructure; investment in these areas can be considered an investment in physical capital.
- Aging economies are not interested in attracting businesses and promoting employment. For instance, they are not likely to lower corporate income tax rates and minimum wages as long as the burden imposed on businesses is not transferred to elderly consumers.

This book reports the statistical analysis of these hypotheses using Japanese and US data to derive implications applicable to other countries.

1.2 Change in Policy Preferences Throughout Voters' Life Cycle

Figure 1.2 depicts the results of a public opinion poll conducted by the Cabinet Office in Japan to understand people's preferences regarding the government's actions. The poll included questions for which multiple answers from different options were permitted. Figure 1.2 shows that many respondents favored policies that protected

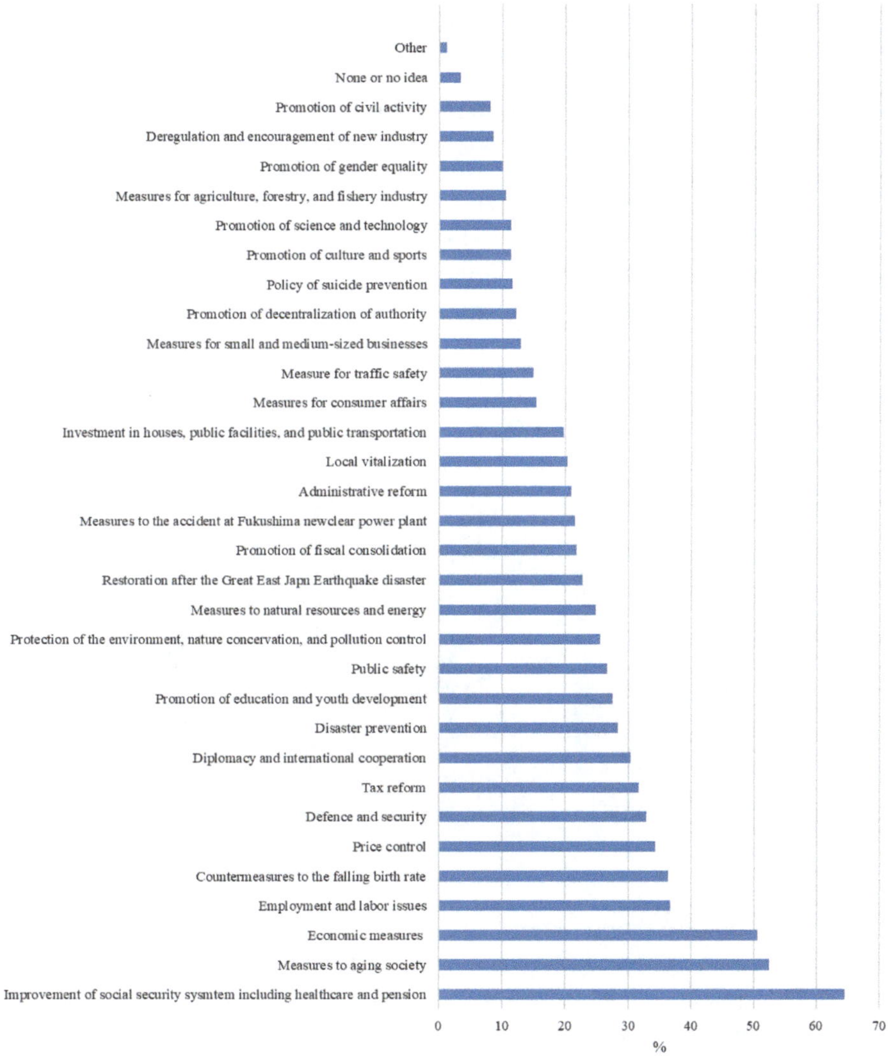

Fig. 1.2 Results of a public opinion poll on desired governmental activities. *Data* Cabinet Office Japan, National Opinion Poll on People's Life, 2018

them from various risks, including those related to pension, healthcare, and unemployment measures. For instance, 64.6% of all the respondents expressed concern regarding the sustainability of the social security system, including aspects such as public pension and healthcare. Further, remarkably, 52.4% of the respondents were concerned about problems caused by population aging. These two aspects may be closely related. Individuals who are anxious about the sustainability or enhancement of the healthcare and pension systems may fear that population aging can seriously threaten these systems.

In this context, a preliminary step is to examine whether people's desires regarding governmental policies vary with age. Figure 1.3 depicts the fractions of respondents in different generations who recorded chosen options. It is noted that two options—"Improvement of social security system including healthcare and pension" and "Measures to aging society"—were chosen by a large proportion in each generation. However, the figure reveals variations with age, as well. The positive response to the two aforementioned options is stronger among older generations, except the oldest age group. This pattern contrasts with the responses to some other options, "Promotion of education and youth development" and "Investment in houses, public facilities, and public transportation." Moreover, the support for these two options is observed to decline with age.

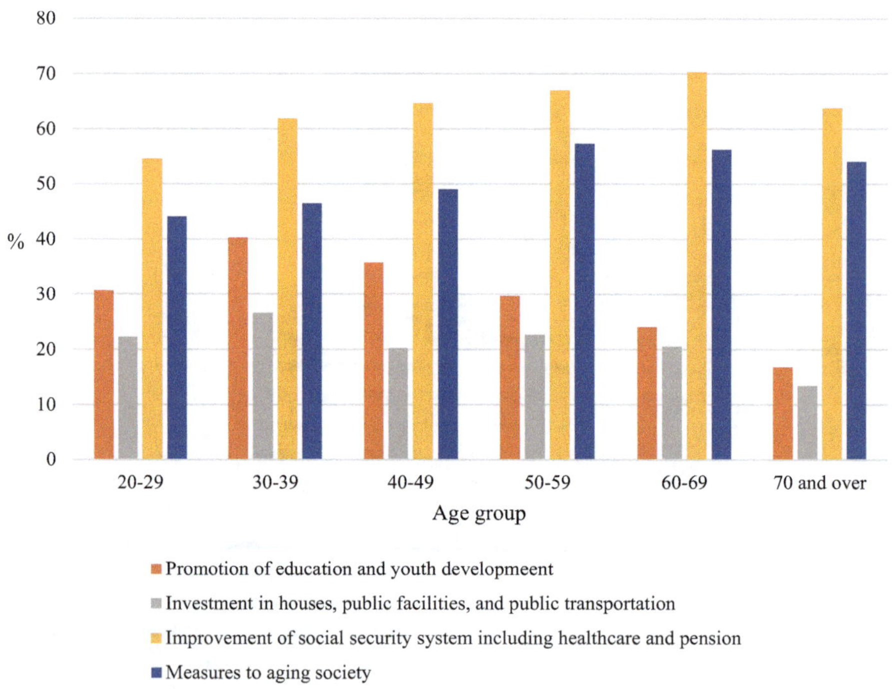

Fig. 1.3 Fractions of respondents in different generations who recorded selective options. *Data* Cabinet Office Japan, National Opinion Poll on People's Life, 2018

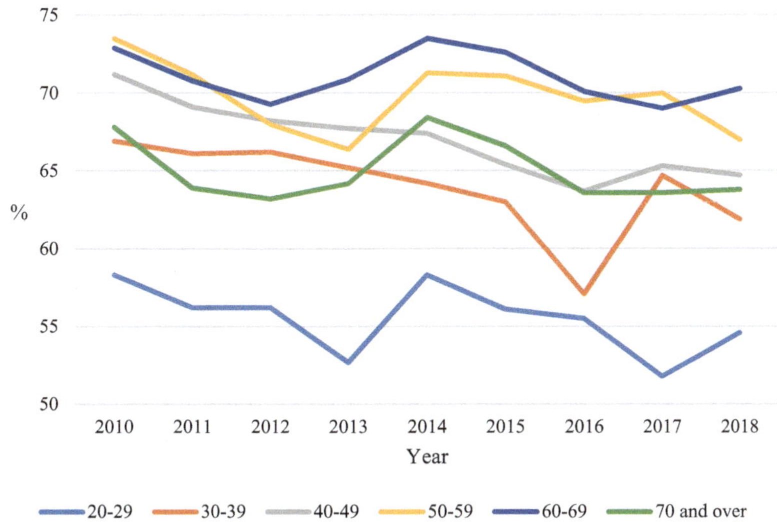

Fig. 1.4 Proportion of respondents who selected the option "Improvement of the social security system including healthcare and pension". *Data* Cabinet Office Japan, National Opinion Poll on People's Life, 2018

Do these facts reflect age effects, rather than cohort effects? Figs. 1.4, 1.5, 1.6 and 1.7 show the fractions of each age group that favored four policies (social security policy, measures to aging society, public investment program, and promotion of education) in different years. Throughout these recorded years, people in their 50s and 60s expressed increasing anxiety about the social security system and the aging society. The people in their 30s have the most interest in public investment in infrastructure, whereas those aged 70 years and over have the least interest. Furthermore, people aged 60 years and over generally do not care about the education of younger generations. Therefore, policy preferences are strongly associated with age.

In summary, population aging causes an increase in the share of constituents who prefer policies that ensure the welfare of the elderly to those that benefit working-age constituents. A society with an increasing elderly population may provide low levels of political support to spending on education and investment. Our empirical analysis aims to examine whether the incumbent politician, or government, will postpone investment in human and physical capital to be reelected in such a society. This action will not hurt the elderly, but hinder growth in the economy.

1.3 Scope of the Analysis

Based on the conjecture presented in Sect. 1.1, we conduct an empirical analysis to investigate how an aging population affects the selection of governmental policies

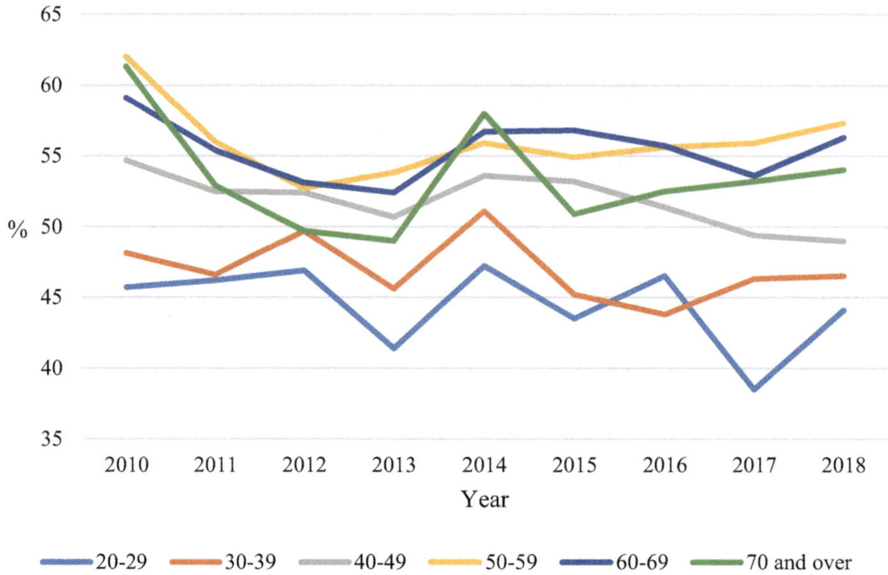

Fig. 1.5 Proportion of respondents who selected the option "Measures to aging society". *Data* Cabinet Office Japan, National Opinion Poll on People's Life, 2018

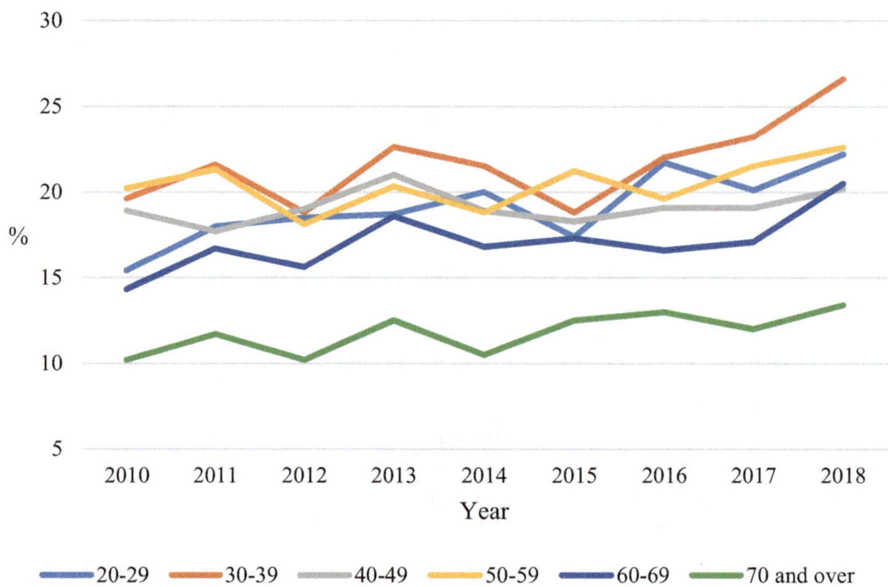

Fig. 1.6 Proportion of respondents who selected the option "Investment in houses, public facilities, and public transportation". *Data* Cabinet Office Japan, National Opinion Poll on People's Life, 2018

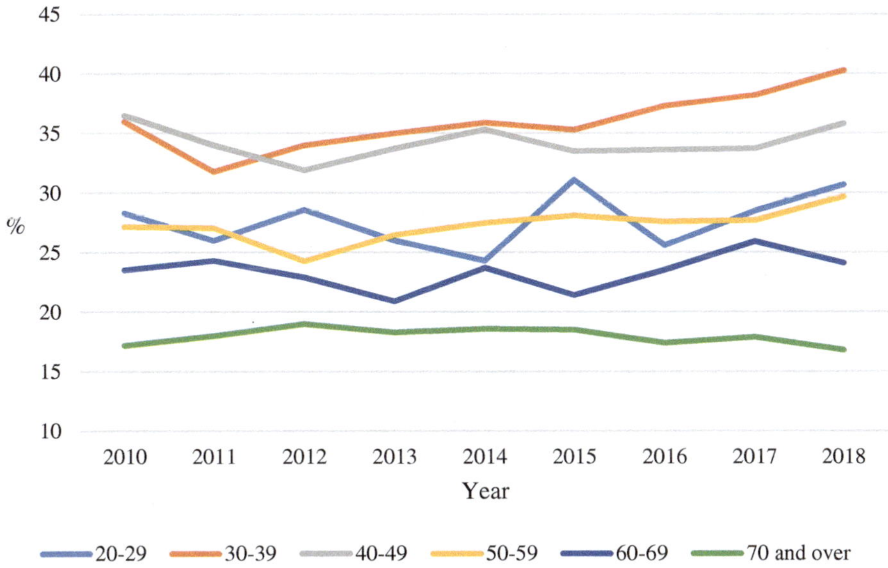

Fig. 1.7 Proportion of respondents who selected the option "Promotion of education and youth development". *Data* Cabinet Office Japan, National Opinion Poll on People's Life, 2018

by focusing on whether a regional government invests less in human and physical capital and implements a high corporate income tax and high minimum wages as long as these policies do not harm the elderly. If the elderly do not send their children to school, drive little, or do not expect to live sufficiently long to benefit from the infrastructure, then we expect them not to support governmental spending on education and infrastructure. However, since several other factors can affect this preference, the analysis outcome is not obvious at this stage.

As stated by Poterba (1998), the elderly may be altruistic, particularly toward their children, and, hence, support spending on education due to the help it affords their children. Second, the elderly may favor housing prices–related spending even when it does not directly benefit them. Consider the case in which the elderly are homeowners and understand that spending $1,000 on education increases demand for living in a particular area and, thereby, increases the area's housing value by $1,500. Then, the elderly will support spending on education even when they do not care about the young and do not need education. By supporting education, they can make a profit of $1,500–$1,000.

The third reason may be externalities. Educated workers may increase the productivity of the elderly who are working. Alternatively, educated younger people may value the consumption of public goods valued by the elderly, as well, or favor the businesses with fixed costs valued by the elderly (e.g., concerts). Such patterns may encourage the elderly to support spending on education.

A fourth reason is related to the labor supply. Suppose that the elderly do not work but consume labor-intensive services, such as nursing care, provided by young

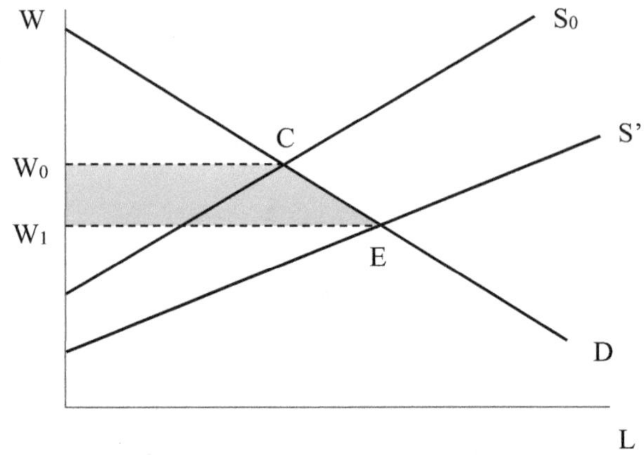

Fig. 1.8 Benefits to the elderly from an increase in labor supply

workers. For example, if the city offers good education facilities, high taxes will be levied; however, it will house large numbers of young workers, which will lower wages for the services required by the elderly. This is described in detail in a theoretical model depicted in Appendix A.1.

This is illustrated in Fig. 1.8, as well. Let the elderly's demand for services provided by young workers be the downward sloping line *D*. The initial supply of young workers is S_0. An increase in education spending increases the labor supply of young workers to S'. This increase results from an increase in young families who value their children's education, the migration of young people, or the productivity of educated workers. An increase in labor supply reduces the labor wage. In Fig. 1.8, the rise in consumer surplus to the elderly is indicated by the shaded area, with vertices w_0CEw_1. If this area exceeds the cost to the elderly of providing improved education, then the elderly will support this spending even when they are not altruistic nor have any children or grandchildren to benefit from it.

A similar analysis considers the taxes or fees imposed on young workers. Suppose that the tax is paid by a worker who writes a check to the government. An increase in the tax will directly benefit the elderly by giving them money. However, it will dissuade the young from moving to the locality, or from working, and thereby cause a rise in wages.

In Fig. 1.9, the initial labor supply by young workers is S_0. A fixed tax T per young worker shifts S_0 upward by T and reduces the labor supply for a given wage to S'. The wage increases from w_0 to w_1, and the quantity of labor supply decreases from L_0 to L_1. At labor quantity L_1, each worker gets w_1, pays the government a fixed tax BC, and has the take-home pay of A. The supply curve S_0 indicates the number of people who want to work for a given take-home pay. At a tax of BC per worker, the total tax revenue, which is assumed to go to the elderly, is indicated by the area with vertices w_1BCA. However, because of the reduced labor supply and higher wages, the consumer surplus enjoyed by the elderly declines from the triangular area with vertices w_0EF to that with vertices w_1BF. In other words, the consumer surplus declines by the trapezoidal area with vertices w_1BEw_0. Therefore,

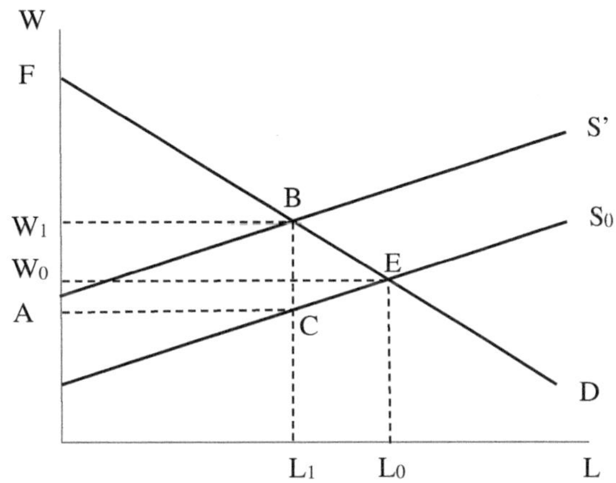

Fig. 1.9 Tax benefits and costs to young workers

if this area exceeds the tax revenue, the elderly will suffer from a tax imposed on young workers. Otherwise, the elderly will benefit from the tax.

Finally, the elderly may have poor cognition and, thereby, make mistakes in voting. Therefore, the theory does not provide an unambiguous answer to whether the elderly will oppose certain types of spending. To clarify this aspect, empirical work is required, which is described in this book.

1.4 Empirical Issues

To estimate the effects of an aged population on policy selection, we used data from Japanese prefectures for three reasons. First, although many countries are encountering population aging–related issues, Japan is aging at a faster rate than most other countries. Second, a wealth of economic, demographic, and political data across long periods is available at the prefecture level. Third, the results obtained by using the data from 47 Japanese prefectures can be compared with the patterns observed in 50 US states since prefectures are positioned in the middle of the vertical hierarchy in the Japanese public sector similar to states in the US. Further, prefectures can establish tax rates and allocate budgets. However, the central government has some control over these prefecture-level activities. For instance, it legislates the Local Tax Act, which specifies the standard rates of taxation by local governments. These features of federalism that prevail in Japan and are partly shared by the US enable us to compare trends in the two countries, despite their different degrees of population aging.

In the regression analysis, we considered expenditure on roads, infrastructure, welfare, and education; the local corporate income tax revenue; and minimum wages as dependent variables. We were particularly interested in the estimated coefficient of the fraction of the population aged 65 years and over. To control political aspects, we used data on the political characteristics of governors and legislators and on the timing of elections. In this manner, we determined the extent to which an aged population influences a regional government to favor pro-elderly spending in Japan and the US.

The high pro-elderly expenditure by a local government may induce in-migration among the elderly, or "voting by their feet," as described by Tiebout (1956). In this case, the ratio of elderly citizens becomes endogenous, which can bias estimation results. To address this endogeneity problem, we applied the instrumental variable regression method.

1.5 Literature Review

To date, many studies have investigated population aging's impact on economy, including Schultz (1992) and Wise (1992) who provided deep and extensive considerations of aging society's evolving concerns about retirement, pension, and health. Casamatta and Batté (2016) reviewed developments in the literature on population aging and its effect on the political economy. They conducted a survey on population aging's impact on education, pension, healthcare (including long-term care), and environmental protection.

The impact of demographic aging can be classified according to the policy fields on which the public expenditure is spent. Some scholars examined how the aging of population affects public expenditure on education. The empirical analysis by Miller (1996) is based on the interest group model of public spending. This model considers the elderly an interest group which opposes high spending on public education and reveals that counties in Texas, US, with large elderly populations, spend less on education. Poterba (1997, 1998) used US state panel data for the fiscal years 1961, 1971, 1981, and 1991. The study's results associated an increase in the proportion of elderly residents in a jurisdiction with a significant reduction in governmental per-child spending on K-12 education. The variation in the size of the school-age population does not cause proportionate changes in education spending, which the author considers sufficient support for models of generational competition in public-sector resource allocation. In Chap. 5, we discuss an empirical analysis using state-level US panel data and perform instrumental variable regression, unlike Poterba (1997, 1998), by considering an individual's choice of a place to live to be endogenous and be influenced by the state government's budget allocation.

A more recent work by Harris et al. (2001) examined the impact of an aging population on public education spending in the US. The authors' analysis of public school district data revealed only a modest negative effect of the elderly on education spending. However, they confirmed that an increasing proportion of the elderly tends to depress state-level spending on education. The authors considered these results to be based on the belief that only local spending is capitalized into house values in comparison with the spending that generates statewide benefits.

Further, demographic change may result in a cutback in government-financed investment. Jäger and Schmidt (2016) examined varying individual returns from public expenditure throughout the life cycle. Senior citizens may discount future benefits more heavily than working-age individuals. Using the panel data of 19 OECD countries for the period between 1971 and 2007, the authors further found a negative impact of population aging on public investment.

Many studies used survey data to analyze the impact of demographic change on the electorate's dominant opinion. Sørensen (2013) addressed issues of voters' life-cycle public spending preferences using the repeated cross-sectional survey data for 22 countries. His results indicate that elderly people prefer less education spending and more healthcare and pension spending. Further, respondents' preferences appeared largely unrelated to their left–right party choices.

Therefore, the existing empirical literature generally indicates the negative impact of an increased proportion of the elderly population on public expenditure on education and investment. Our analysis adds new insights to the earlier works, as follows. First, our statistical estimation included variables that capture regions' institutional and political features as control variables. These variables, including governors' and legislators' political traits and reelection motives, helped clarify the non-demographic effects of political processes. Second, the results of our regression analysis on Japanese and US data helped derive implications for countries at different population aging stages.

The budget allocation by a government intending to gain the elderly's political support may be associated with low future growth rates. Gonzalez-Eiras and Niepelt (2012) analyzed the short- and long-term effects of demographic aging on per capita growth. Their overlapping generations model that incorporated political processes predicted that retirement age increases and per capita growth accelerates in response to demographic changes. A fixed retirement age can cause the stagnation of per capita growth rate due to a surge in social security transfers and crowding out of public investment, which can increase productivity.

1.6 Roadmap to This Book

Sect. 1.2 examined the relationship between population aging and public opinion formation; the older generation is very concerned about pensions, healthcare, and elderly welfare, whereas younger generations strongly favor the promotion of education and public investment programs. The politico-economic perspective indicates that an aging society likely adopts policies with short-lived benefits. The following chapters investigate demography's effects on various programs implemented by regional governments.

Chapter 2 examines how the proportion of elderly individuals in Japanese prefectures affects prefectural governments' infrastructure-related expenditure. The results show that population aging negatively affects investment in public capital. Further, the ratio of the number of ruling party members to the total number of elected parliamentarians in each prefecture is associated with governmental spending on public capital investment. Parliamentarians belonging to the ruling party are considered to play an important role in obtaining financial resources from the central government. However, the effect of party affiliation is dominated by the effect of population aging. Many earlier analyses have shown that, in Japan, public capital stock is excessive in rural areas but inefficiently insufficient in urban areas. Although population aging may increase efficiency in the short term by mitigating excess capital stock in rural areas, it may enhance the capital stock shortage in urban areas in the long term.

The impact of population aging on governmental spending on education, elderly welfare, social welfare, and public assistance is examined using Japanese prefectural data and by controlling changes in various systems in Chap. 3. Aging may positively affect public transfers that do not produce long-term benefits. The estimation results obtained using the instrumental variable method confirm that a large proportion of the population aged 65 years and over is associated with an increase in social welfare, public assistance, and elderly welfare expenditures on a per capita basis. On the other hand, prefectures whose populations have large proportions of people aged 65 years and over spend less on education. Therefore, an increase in the elderly population ratio corresponds to an increase in expenditure that directly benefits the elderly and a suppression of expenditure that generates long-term benefits but does not directly benefit the elderly.

Chapter 4 examines how population aging affects the taxation of corporate income in Japan. Since the elderly's participation in the labor force is significantly low, elderly voters may prefer the imposition of high corporate income taxes over increases in other taxes. On the other hand, if low corporate tax rates stimulate corporate activities and increase the value of local real estate and local companies, the elderly—who own relatively large volumes of real estate and financial assets in Japan—may support corporate tax rate reductions. Moreover, Japan has a mechanism in place to adjust shortages in local general financial resources using local allocation tax grants. This ensures that a decrease in local governments' corporate tax revenue does not immediately cause a reduction in expenditures for the elderly. Using this measure, even reducing corporate tax rates, local governments can maintain their spending for the elderly. In such circumstances, there is only a weak association between reductions in corporate tax revenue and reductions in the spending for the elderly. The estimation results presented in Chap. 4 are consistent with these views. They reveal that the elderly do not oppose reductions in corporate taxes.

Chapter 5 presents estimation results for public expenditure data at the state level in the US. The proportion of spending on education is shown to decline with an increase in the elderly's share in the population. However, the proportion of spending on highways is not significantly affected by the elderly's population share in US states, in contrast to the results on the spending on public capital investment in Japan. This may be because of citizens' mobility across US states. Young citizens who may leave their current area of residence may not like to invest in regional infrastructure. In other words, the decreased political influence of younger citizens does not cause a decline in regional investment in highways. Further, an increase in the proportion of Hispanic or Latino people is associated with an increase in educational spending, which suggests that racial composition is another factor affecting regional budget allocation.

Chapter 5 reports the results of another regression, which considered educational spending per pupil a dependent variable, as suggested by Poterba (1997, 1998). The panel data analysis reveals that the real capital outlay per pupil decreases with an increase in the share of elderly citizens; however, this effect is not found for the real current spending per pupil. The application of the instrumental variable regression method shows the disappearance of the effect on real capital outlay per pupil. Hence, interstate migration, which leaves regional governments' expenditure structure unchanged, may mitigate the impact of an increase in the elderly's political power.

Chapter 6 examines how the proportion of the elderly population in a state affects the state government's policy on taxation and intervention in the labor market. One of our hypotheses, which is presented in Sect. 1.1, is that aging economies pay less attention to lowering the tax burden on local businesses. However, the elderly may favor a low minimum wage because an increase in the minimum wage may increase the prices of goods and services. Moreover, if the tax burden on local businesses is transferred to price increases, then the elderly may prefer a low corporate income tax rate.

Regression analysis shows that a larger proportion of the elderly population in a state is associated with a decline in the state's revenue share from corporate income tax. This indicates that elderly citizens are concerned about the indirect effects of the increase in tax burden on firms. On the other hand, our estimates show no significant impact of the elderly population's interests on the minimum wage. The labor market structure may explain this result. If employers have monopsony power in the local labor market for low-wage workers, the minimum wage may have little effect on the prices of goods and services that are produced by low-wage labor and consumed by the elderly, such as nursing care. In this case, the elderly may not strongly oppose the local government's establishment of a higher minimum wage.

Chapter 7 considers motives of voting other than driven by self-interest; altruism, intergenerational bargaining, and expressive voting. The survey of previous empirical works shows that these factors to some extent affect the outcome of increased elderly population. Moreover, Chap. 7 discusses applicability of our results to countries other than Japan and the US. It concludes that the effects of population aging will differ across countries. Residential mobility affects altruistic motives of parents and children. Also, impacts of population aging vary across genders and races. That in turn affects political support for different policies.

In conclusion, our hypotheses, which are based on political economy, are mostly supported empirically, although the policy fields for which demographic effects are apparent differ between Japan and the US. In Japan, the impact of the aging population on the government's investment in physical capital is significant, whereas, in the US, it does not have a strong effect on public spending on highways. This difference may be explained by the difference in individuals' mobility across regions, which may come from diverging degrees of fiscal decentralization; the different policy choices offered by regional governments can motivate citizens to vote by their feet. This enables the political minority to escape from the exercise of power by the political majority. Therefore, when predicting the impacts of population aging, one should consider the institutional elements of any country.

Appendix 1 Community Spending and Taxation and Young Worker Migration

Consider a community in which the elderly form the major proportion of the population. The numbers of young people and the elderly in the community are n_y and n_o, respectively. Suppose that all other communities comprise young residents alone. The young are mobile, whereas the elderly are not.

The community with the elderly population has a fixed budget (B). The budget is allocated between a public good that is valued only by the elderly (G_o) and a public good that is valued only by the young$\left(G_y\right)$. The budget constraint for the community is $G_y + G_o = \mathrm{B}$.

An old person's utility is $U\left(G_o, \frac{n_y}{n_o}\right)$, which represents an increasing and concave function of each argument. A high ratio of young to old people benefits the elderly, because young labor can provide much service to each old person.

If a young person lives and works in a community having elderly persons, his or her utility depends on their wage, $w\left(\frac{n_y}{n_o}\right)$. The wage decreases with an increase in the number of young people per old person, because this increases the supply of labor relative to demand. A young person's utility is $w\left(\frac{n_y}{n_o}\right) + \mathrm{V}\left(G_y\right)$, where V is an increasing function of G_y.

Let a young person have the reservation utility $\acute{V}$ which is attained in communities inhabited by young people alone. When G_y is in equilibrium, the number of young people per old person is implicitly given by

$$w\left(\frac{n_y}{n_o}\right) + \mathrm{V}(G_y) = \acute{V}. \tag{A.1}$$

Equation (A.1) determines the number of young people per old person as a function of G_y, that is, $\frac{n_y}{n_o} = \varphi\left(G_y\right)$ with $\varphi' = -\frac{V'}{w'} > 0$.

Hence, by providing some G_y, the elderly can attract more young people and increase $\frac{n_y}{n_o}$ and, thereby, the elderly's utility. On the other hand, they are constrained by the community's fixed budget, such that an increase in G_y causes a decrease in G_o. The value of G_o most preferred by the elderly satisfies

$$\frac{\partial U}{\partial G_o} - \frac{\partial U}{\partial\left(\frac{n_y}{n_o}\right)}\varphi\prime = 0. \tag{A.2}$$

Therefore, the elderly who rely on the young for services, such as healthcare and transportation, provide some $G_y = B - G_o$ to attract young labor. If we interpret G_y as the quality of education in the community, the higher the G_y, the higher the future output.

We can perform a similar analysis for fixed quantities of G_y and G_o with the community's decisions on the taxes to be levied on the young and old. The community's budget constraint is $G_y + G_o = n_y T_y + n_o T_o$ where T_y and T_o are the taxes imposed on a young and an old person, respectively.

A young person's utility is $w\left(\frac{n_y}{n_o}\right) - T_y + \mathrm{V}(G_y)$. The condition corresponding to Eq. (A.1) determines the number of young people per old person as a function of T_y, $\frac{n_y}{n_o} = \rho(T_y)$, with $\rho' = \frac{1}{w'} < 0$. Then, the value of T_o to maximize the elderly's utility, which is defined as $U\left(G_o, \frac{n_y}{n_o}\right) - T_o$, is

$$-\frac{\partial U}{\partial\left(\frac{n_y}{n_o}\right)}\rho'\frac{n_o}{n_y} - 1 = 0. \tag{A.3}$$

Hence, the elderly will pay some taxes to induce the young to live in their community.

References

Black D (1948) On the rationale of group decision-making. J Pol Econ 56(1):23–34
Casamatta G, Batté L (2016) The political economy of population aging. In: Piggott J, Woodland A (eds) Handbook of the economics of population aging, 1A. North-Holland, pp 381–444
Downs A (1957) An economic theory of democracy. Harper Collins, New York
Gonzalez M, Niepelt D (2012) Ageing, government budgets, retirement, and growth. Eur Econ Rev 56(1):97–115
Harris AR, Evans WN, Schwab RM (2001) Education spending in an aging America. J Public Econ 81(3):449–472
Jäger P, Schmidt T (2016) The political economy of public investment when population is aging: a panel cointegration analysis. Eur J Pol Econ 43:145–158
Miller C (1996) Demographics and spending for public education: a test of interest group influence. Econ Educ Rev 15(2):175–185
Poterba JM (1997) Demographic structure and the political economy of public education. J Pol Anal Manage 16(1):48–66
Poterba JM (1998) Demographic change, intergenerational linkages, and public education. Am Econ Rev 88(2):315–320
Schulz JH (1992) The economics of aging. Praeger Pub Text. In: Schulz JH (ed) (1998) Eijingu no keizaigaku (trans. Sato R, Sagaza H, Sato Y). Keio Shobo, Tokyo
Sørensen RJ (2013) Does aging affect preferences for welfare spending? A study of peoples' spending preferences in 22 countries, 1985–2006. Eur J Pol Econ 29:259–271
Tiebout CM (1956) A pure theory of local expenditures. J Pol Econ 64(5):416–424
Wise DA (ed) (1992) Topics of the economics of aging. University of Chicago Press, Chicago

Chapter 2
Infrastructure-Related Expenditures and Population Aging in Japan

2.1 Introduction

How does aging affect government activities? This chapter empirically analyzes this concern by focusing on public spending, such as public investment. Some studies have confirmed that aging negatively impacts public spending that brings long-term benefits for us. Jäger and Schmidt (2016) used OECD data to analyze how aging affects public capital investments. Their study shows that aging negatively impacts investments of public capital. To the best of the authors' knowledge, no study has yet analyzed the effects of aging on public capital investments in Japan. Instead, studies on public capital investments in Japan have focused on their productivity and efficiency. Some of these studies show that Japan's public capital investment was performed inefficiently and allocated unfairly across regions. This biased regional allocation is attributed to political processes. Jäger and Schmidt (2016) did not analyze bias in the distribution of public capital among regions. In this chapter, we analyze public capital investment from the perspectives of both aging and regional allocation.

Figure 2.1 is a map that indicates per capita infrastructure-related expenditure for each district, whereas Fig. 2.2 shows the ratio of ruling-party politicians against all parliamentarians elected from each district. The data are average figures from 1995 to 2008. Tokyo, Kanagawa, Osaka, and Aichi are the top four prefectures in Japan regarding population size. Figure 2.1 shows that per capita infrastructure-related expenditure is relatively low in these districts compared to other areas, while per capita infrastructure-related expenditure is relatively high in regional areas. Figure 2.2 shows that many of the areas with a high ratio of ruling party parliamentarians are regional districts. For example, the ratio is high in the Hokuriku, Shikoku, and Chugoku areas. In Kyushu, the southern part also has a high proportion of parliamentarians belonging to the ruling party. The data cover the period during which the Liberal Democratic Party (LDP) served as the governing party. The LDP had a politically advantageous position in these districts during that period. Figure 2.3 shows a scatter diagram for Figs. 2.1 and 2.2. The diagram plots the average values for

K. Terai et al., *The Political Economy of Population Aging*, Advances in Japanese Business and Economics 30, https://doi.org/10.1007/978-981-16-5536-4_2

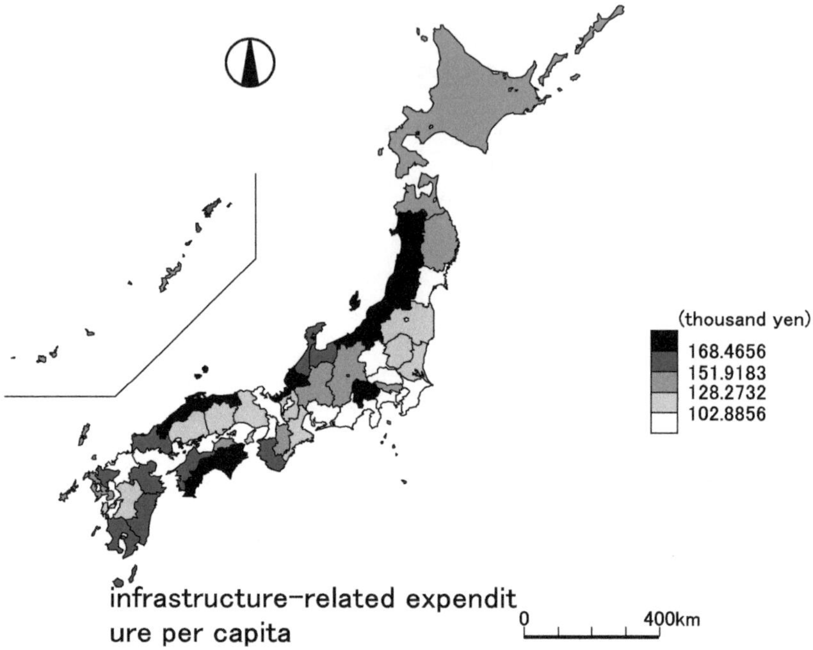

Fig. 2.1 Infrastructure-related expenditures at the prefectural level

the sample period, in which no regime change took place, to clarify the relationship between the ratio of ruling party parliamentarians and public capital investments. Figure 2.3 indicates a positive correlation between the ratio of ruling party politicians among regionally elected parliamentarians and per capita infrastructure-related expenditure, even though the slope may not be steep. The high ratio of ruling party parliamentarians in regional districts is related to the level of per capita infrastructure-related expenditure in the context of Japan's public capital spending. This point should also be considered when investigating this issue. This chapter analyzes Japan's public capital spending from two perspectives: the impact of population aging and the ratio of ruling party parliamentarians.

The remainder of this chapter is organized as follows. Section 2.2 presents the literature review, while Sect. 2.3 describes the relationship between revenue and expenditure at Japan's central and local government levels, addressing the influence of ruling party parliamentarians. Section 2.4 discusses the data and the method of estimation. The results of the estimation and influence of population aging and ruling party parliamentarians on public capital spending are presented in Sect. 2.5. Section 2.6 concludes the chapter.

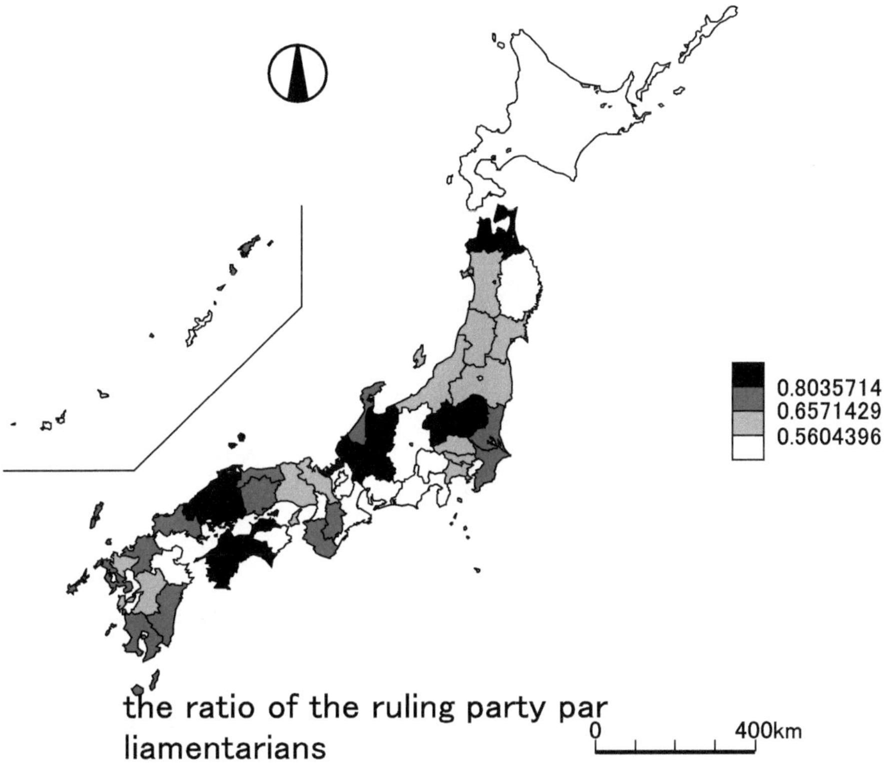

Fig. 2.2 The ratio of the ruling party parliamentarians at the prefectural level

2.2 Literature

As a percentage of GDP, public capital spending is declining in major industrialized countries, a phenomenon that several studies attribute to "fiscal pressure" (Mehrotra and Välilä 2006; Vuchelen and Caekelbergh 2010; Bacchiocchi et al. 2011). In contrast, Jäger and Schmidt's (2016) empirical analysis attributes the decline in public capital spending to population aging. They conducted a panel cointegration analysis using data from 1970 to 2007 in 13 OCED countries and found a negative correlation between aging and public capital investment. In addition, based on the Granger causality test, they found a one-way causality from aging to public capital investment. Regarding empirical analyses of public capital using Japanese data, Yoshino and Nakano (1994), Asako et al. (1994), Doi (1998), and Yamano and Ohkawara (2000) investigated the efficiency of public capital and regional allocations. Estimation results of these studies found a bias in the allocation of public capital stock among regions. Conversely, Kondoh (2008) analyzed the regional distribution of public capital investment from a political economy perspective. It used prefectural panel data from 1980 to 2003 and the estimation results showed that the number of

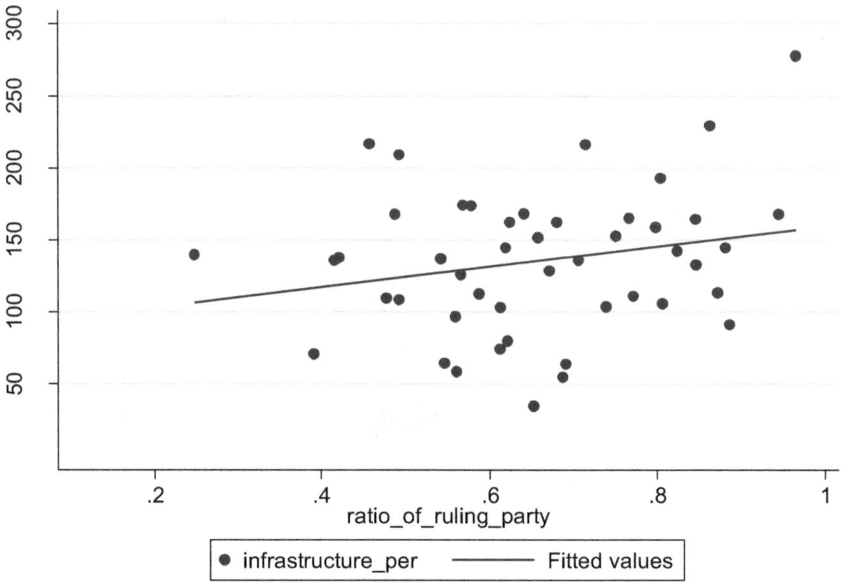

Fig. 2.3 Relations between infrastructure-related expenditures and ratio of ruling party parliamentarians

LDP members has a significant positive impact on public capital expenditure, particularly on public investments, such as industrial infrastructure, infrastructures for life, agriculture, forestry and fisheries, and conservation of national land.

This chapter analyzes how aging affects public capital investment. Some studies have investigated public capital investments from the viewpoint of optimal scale and efficiency. Arrow and Kurz (1970) theoretically analyzed the optimal size of public capital, and Aschauer (1989) analyzed the returns and productivity of public capital by estimating production functions that consider public capital. Iwamoto (1990) estimated the returns and productivity of public capital in Japan. Subsequently, several studies in Japan have analyzed the productivity and efficiency of public capital (Asako et al. 1994; Yoshino and Nakano 1994; Mitsui and Ohta 1995; Doi 1998; Yamano and Ohkawara 2000; Hayashi 2009).

2.3 Revenue and Expenditure at the Central and Local Government Levels

Section 2.3 provides an overview of the structure of revenue and expenditure at Japan's central and local government levels to discuss the influence of ruling party parliamentarians on public capital investments. Figure 2.4 shows the structure of revenue and expenditure at the central government and local government levels.

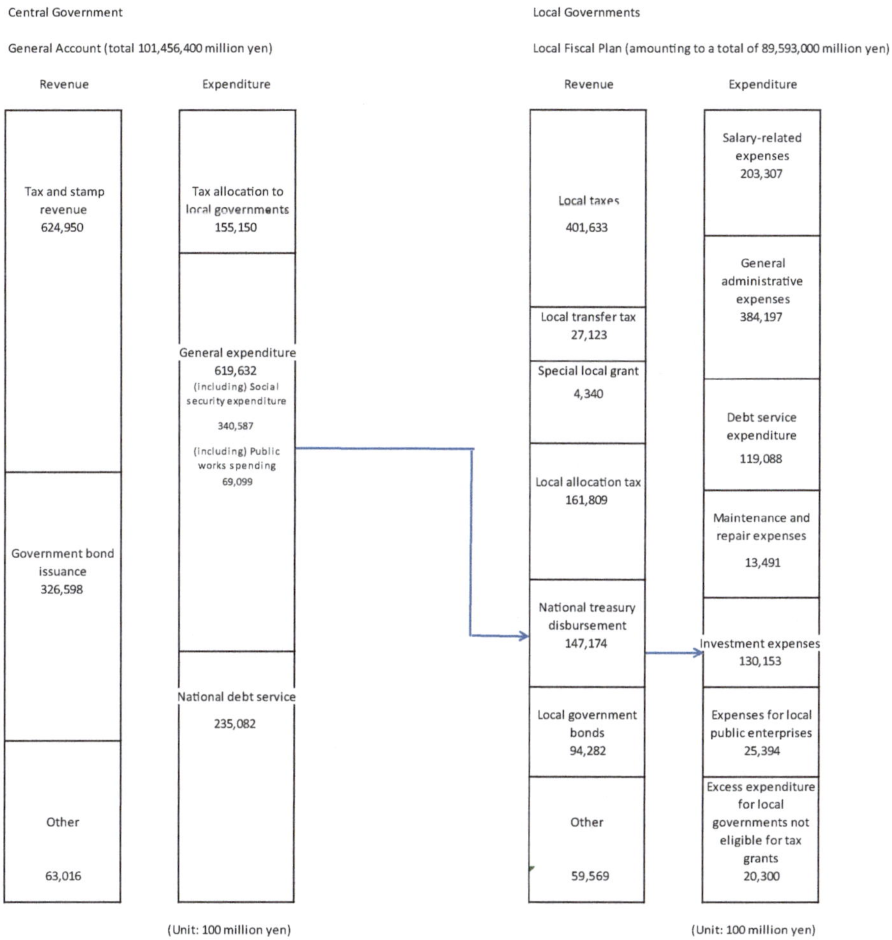

Fig. 2.4 Revenue and expenditure of the central government and local governments

According to fiscal 2019 data, the central government's revenue and expenditure was 101.4 trillion yen, compared with 89.5 trillion yen for local governments. The size of revenue and expenditure did not show a significant difference between the central government and local governments, but the revenue breakdown was different. Local governments generated 40.1 trillion yen in tax revenue, which comprised only about 45% of their overall revenue. Expenditure that exceeds local tax revenue is financed by fiscal transfers from the central government. Subsidies of tax allocated to local governments and national treasury disbursements play a major part in fiscal transfers from the central government to local governments. The fiscal conditions of local governments differ from one another, depending on their economy and industrial structure. Subsidies of tax allocated to local governments are designed to narrow the disparities in public services provided to residents. After determining the revenue deficit of each local government, the central government allocates subsidies to local governments based on a formula prescribed by the Local Allocation Tax Act.[1] There are no restrictions on the use of money allocated to local governments under this system.

Contrarily, "national treasury disbursements" include contributions, commission fees, incentives for specific measures, or financial assistance that the central government pays to local governments. National treasury disbursements primarily comprise national treasury contributions, deposits, and subsidies. Expenditures for public works, such as social capital construction projects, are often transferred from the central government to local governments in the form of national treasury subsidies. National treasury disbursements, unlike tax allocations, restrict how the grants-in-aid are used. National treasury disbursements are a source of revenue earmarked for specific items.

We consider the relationship between parliamentarians and public capital investments. In Japan, in several instances, public works projects are financed through the transfer of funds from the central government to local governments in the form of national treasury disbursements. Ultimately, the parliament determines national treasury disbursements to local districts, which is where parliamentarians may intervene. In Japan, the LDP has served as the ruling party for a long time. There were many LDP parliamentarians known as *doro zoku* (road-tribe lawmakers) who exerted a strong influence on public works spending. Figure 2.3 shows data for the period during which the LDP served as the ruling party. The figure demonstrates that areas with a high ratio of members of the ruling party among locally-elected parliamentarians also had high public capital spending per capita (for the period in Fig. 2.3, the ruling party was the LDP). Therefore, while analyzing public capital investments in Japan, it is necessary to consider the political process underlying national treasury disbursements involving parliamentarians elected from regional districts.

2.4 Data and Estimation Strategy

We used the primary prefectural data in the "Social and Demographic System" provided by the Ministry of Internal Affairs and Communications. This dataset is long-term panel data. A panel analysis that considers individual effects in regions is possible. The dataset contains general account budgets for each local government, total production, income, and population by age in each prefecture. In addition to the

[1] The following is the prescribed formula: The amount of ordinary tax allocation for each entity = the amount of basic fiscal demand—the basic fiscal revenue = the revenue shortfall. It is based on two other formulas: the amount of basic fiscal demand = the unit cost (legally prescribed) × the measurement unit (the population according to national census data, etc.) × correction coefficients (adjustments for regions with cold temperatures, etc.); the basic fiscal revenue = the standard tax revenue projection × the basic taxation rate (75%). The unit cost used in calculating the basic fiscal demand is the amount of general fund per measurement unit that local governments need for standard administrative tasks.

impact of aging on public capital investment, this chapter also considers regional bias in public capital investment through political processes. We use information such as the electoral districts and political parties of the Diet members of the "Parliamentary Handbook" to analyze political processes in regions. The sample period is 1996–2014. In Japan, the electoral system changed from a medium-size constituency to a single-seat constituency system in 1996. We use a sample after 1996 to control for the electoral system change. The basic formula for the estimation is as follows:

$$y_{i,t} = \alpha + \beta x_{i,t} + \beta^j X^j_{i,t} + \gamma_i + u_{i,t} \tag{2.1}$$

Here, the explained variable (linfrastructure) is the log of infrastructure-related expenditure per capita. The data in general account expenditures provide each item. We consider civil engineering expenses, road and bridge expenses, and city planning expenses as infrastructure-related expenditures, and we divide them by the total population in each prefecture. In addition, we take a log and use it as the explained variable. γ_i is an individual effect, and $u_{i,t}$ is an error term. Aging is an interest variable in this chapter. We divide the population aged 65 and older by the total population in each prefecture and calculate the log. We use the log as a variable for aging (lratio_aged65). $X^j_{i,t}$ are the control variables. We use the log of the ratio of the population aged below 15 years (lratio_15), the log of per capita gross production (lgdp), and the log of per capita income (lincome) in each prefecture. lgdp is a variable used to control for annual economic conditions, and lincome is a variable to control for the income level in each prefecture. We include the log of the active opening ratio (lactive_open) to control for the job market and conditions in the region. We divide the nominal value by the deflator in each prefecture to calculate real terms in per capita infrastructure-related expenditure, per capita gross domestic product, and per capita income. In addition, we include the log of the fiscal power index (lfiscalpower) in each prefecture. This variable represents the fiscal soundness of each local government.

Furthermore, we include the yearly trend (trend) in the estimation to control for socio and economic trends and a dummy variable in 2008 (year2008) to control for the Lehman shock. From a political economy's point of view, politicians will organize budgets to gain an advantage in their election. Governor elections are important at the local government level. Therefore, we make an election dummy variable (election) that takes the value of 1 when the governor's election is held or is otherwise 0 for each prefecture. We include this variable in the estimation to consider the effects of the governor's election on the local government's budget. In addition, there may have already been operating budgets before the governor's election year. Therefore, we make a lead variable of the governor's election-year (election_F1) and add it to the estimation.

Furthermore, we add a lag variable of the governor's election-year (election_L1) in the estimation. This variable is considered a reward for supporting the candidate after the governor's election. There is a regional bias in public capital investment in Japan, which is a distortion of the political process. We consider these effects on public capital investments. The central government finances most public capital investments in Japan. Regarding the budgets of the central government, the role of parliamentarians, particularly ruling parliamentarians, is significant, and they have the authority to determine budget allocations. The extent to which each prefecture elects the ruling parliamentarians may affect the amount of public capital investment in these regions. Therefore, we calculate the ratio of the ruling party members to the elected total parliamentarians in each prefecture and take a log (lratio_ruling). We include this variable in the estimations.

We conducted a panel analysis that considers individual effects in regions in the estimation (2.1). However, if local governments' budgets affect those aged 65 and older in the region, the estimated coefficient of lratio_aged65 is biased. People aged 65 and older may migrate to municipalities that provide higher ratios for budget items preferred by the elderly. They may leave local governments where the budget ratio they prefer is low. If there are residential movements of the elderly between local governments, the estimation in formula (2.1) has a bias. We used the instrumental variable method to deal with the problem of elderly residential movements driven by budget allocation in local governments. Assuming that the budget ratio by item affects the population ratio for people aged 65 and older, the explanatory variable correlates with the error term. The consistency in the estimation cannot be guaranteed in the panel analysis, even with the fixed-effects model or the random-effect model. In this case, the instrumental variable method can be used to obtain a consistent estimator.

$$x_{i,t} = \tau + \pi z_{i,t} + \epsilon_{i,t} \tag{2.2}$$

Here, z is an instrumental variable. There should not be a correlation with the error term $u_{i,t}$ in (2.1). In this paper, we use past demographic structures, following Harris, Evans and Schwab (2001), Ladd and Murray (2001). We use the log of the ratio of people aged 65 and above to the total population for the past 10 years (lratio_aged65past) as an instrumental variable. The ratio of people aged 65 and above in the past is correlated with the current aged population ratio (lratio_aged65). Conversely, it is not affected by the current explained variable (linfrastructure). Therefore, the variable lratio_aged65past is considered an appropriate instrumental variable. Table 2.1 shows the descriptive statistics of variables in the estimations. In the estimations, we conducted a regression with the variables, taking the logarithms of infrastructure, ratio_aged65, ratio_aged15, income, gdp, active_opening, fiscalpower, and ratio_ruling.

2.5 Estimation Results

First, we present the estimation results in the pooled data. Column (1) in Table 2.2 shows the results. The ratio of those aged 65 and above positively affects per capita infrastructure-related expenditure, contrary to our hypothesis. As for the political economy variables, the current and previous period dummy variables for governor's

Table 2.1 Descriptive statistics

	Mean	Std. Dev.	Min	Max
Dependent variable				
Infrastructure	125.859	61.776	17.101	381.777
Explanatory variables				
ratio_aged65	0.219	0.041	0.105	0.326
ratio_aged15	0.141	0.013	0.108	0.215
income	2595.023	401.980	1818.347	4418.649
gdp	3.415	0.667	2.365	7.189
active open	1.869	0.798	0.516	6.413
fiscalpower	0.459	0.196	0.197	1.405
election	0.253	0.435	0	1
election_L1	0.264	0.441	0	1
election_F1	0.264	0.441	0	1
ratio_ruling	0.695	0.266	0	1
democratic	0.210	0.407	0	1
trend	31.000	5.480	22	40
year2008	0.052	0.223	0	1
Instrument				
ratio_aged65_10yearsago	0.164	0.039	0.073	0.267
Observations	893			

election are not statistically significant. In addition, the variable of the ratio of the ruling party parliamentarians is positive but not statistically significant. In the estimation of column (1) in Table 2.2, the individual effects of the region were not considered. We conducted the random effect model and the fixed-effects model to consider these individual effects in columns (2) and (3) in Table 2.2, respectively. The estimation results in column (2) in Table 2.2 show that the coefficient of the population ratio for people aged 65 and older is positive and statistically significant. The fixed-effects model shows that this aging variable is not statistically significant. The variables of the proportion of people aged below 15 years are not statistically significant in columns (2) and (3) in Table 2.2. Regarding the governor election dummies, the current period dummy is not statistically significant. However, the pre-period dummy (election_F1) of the governor election is positive and statistically significant. This means that infrastructure-related expenditure increases when governor elections draw near. Thus, it supports the theory of the political business cycle in Nordhaus (1975). Furthermore, the after-period dummy (election_L1) of the governor election is negative and statistically significant. This also implies that the political business cycle works. The coefficients of the ratio of ruling party parliamentarians are positive but not statistically significant in columns (2) and (3) in Table 2.2.

Table 2.2 Estimation results

	Pooled OLS	Random Effect	Fixed Effect	IV	IV First stage
	(1)	(2)	(3)	(4)	lratio_aged65
lratio_aged65	0.6407**	0.6976***	0.2394	−0.8226**	
	(0.2487)	(0.2621)	(0.2788)	(0.3748)	
lratio_aged15	0.5697	0.4485	−0.1275	0.3425	0.6387***
	(0.4143)	(0.3702)	(0.4981)	(0.5290)	(0.1288)
lincome	0.7992**	1.0921***	1.7284***	1.1217*	−0.3458**
	(0.3896)	(0.3915)	(0.5075)	(0.6042)	(0.1352)
lgdp	1.1705***	0.2790	−0.3663	0.0523	0.2991*
	(0.3935)	(0.4211)	(0.5820)	(0.6990)	(0.1491)
lratio_ruling	0.0466	0.0082	0.0285	0.0448*	0.0081*
	(0.0287)	(0.0210)	(0.0231)	(0.0247)	(0.0044)
election	−0.0108	−0.0078	−0.0046	−0.0048	−0.0007
	(0.0116)	(0.0088)	(0.0089)	(0.0093)	(0.0024)
election_L1	−0.0204*	−0.0206**	−0.0187**	−0.0207**	−0.0034
	(0.0112)	(0.0090)	(0.0090)	(0.0095)	(0.0018)
election_F1	0.0101	0.0126*	0.0161**	0.0172**	0.0002
	(0.0104)	(0.0074)	(0.0078)	(0.0080)	(0.0019)
lfiscalpower	−1.1615***	−0.5591***	−0.1392*	−0.1656**	−0.0225
	(0.1095)	(0.0958)	(0.0725)	(0.0764)	(0.0167)
lactive_open	0.0852**	0.1208***	0.1313***	0.1142***	−0.0040
	(0.0374)	(0.0246)	(0.0246)	(0.0268)	(0.0046)
trend	−0.0518***	−0.0556***	−0.0513***	−0.0190	0.0049
	(0.0081)	(0.0091)	(0.0109)	(0.0134)	(0.0030)
year2008	0.0063	−0.0510***	−0.0812***	−0.0836***	−0.0103***
	(0.0168)	(0.0146)	(0.0161)	(0.0163)	(0.0020)
lratio_aged65					0.8439***
_10yearsago					(0.1092)
sigma_u		0.1405	0.4673	0.5425	0.0788
sigma_e		0.1228	0.1229	0.1281	0.0268
rho		0.5666	0.9353	0.9471	0.8962
R-squared	0.8385				
Within		0.8310	0.8500	0.8369	0.9697
Between		0.8351	0.0080	0.2915	0.7745
Overall		0.7366	0.2082	0.0585	0.8666

(continued)

Table 2.2 (continued)

	Pooled OLS	Random Effect	Fixed Effect	IV	IV First stage
	(1)	(2)	(3)	(4)	lratio_aged65
Observations	868	868	868	868	868

Note *p < 0.1; **p < 0.05; ***p < 0.01. Cluster-robust standard errors are shown in parentheses. Column (1) shows the estimation result based on the pooled OLS; column (2) shows the estimation result based on the random effect model, column (3) shows the estimation result based on the fixed-effect model, and column (4) shows the estimation results based on the instrumental variable method

An assumption for the random effect model is that there is no correlation between the explanatory variables and an error term. If there is self-selection due to the choice of residence location, the assumption would be violated. Thus, it seems better to employ a fixed-effects model to control for individual regional effects.

Furthermore, if per capita infrastructure-related expenditure affects the population ratio of people aged 65 and older, even the fixed-effects model that considers individual regional effects has a bias in its estimations. We conducted the instrumental variable method to deal with the endogeneity problem. Column (4) of Table 2.2 shows the estimation results based on the instrumental variable method. The 65-year-old and over-population ratio variables are negative and statistically significant. Therefore, as the region's 65-year-old and over-population ratio increases, the infrastructure-related expenditure per capita will decrease, and aging will result in lower public capital investment. Based on the estimated coefficient, a 1% increase in people aged 65 and older leads to an increase of 82.3% per capita infrastructure-related expenditure. This appears to be a large value. Thus, we consider standard deviations such as the standardized coefficient. We calculated this as $\beta^{s}_{lratio\_aged65} = \beta_{lratio\_aged65} \times \frac{S.D(lratio\_aged65)}{S.D(linfrastrcture)}$. The calculated value is −0.311. Based on this calculation, one standard deviation in the ratio of the elderly decreases the ratio of infrastructure-related expenditure by 0.311. The coefficient for the population ratio aged below 15 years is positive but not statistically significant. The population ratio aged below 15 years does not have a significant effect on the instrumental variable method. Regarding other political economy variables, the current period of the governor's election is not statistically significant. The pre-period dummy of the governor's election is positive, the after-period dummy is negative, and both are statistically significant. This means that infrastructure-related expenditures increase before the governor's election and decrease after it. This result supports the theory of the political business cycle in Japan. Furthermore, the ratio of ruling party parliamentarians has positive and statistically significant effects.[2] This means that a higher ratio of ruling party parliamentarians in regions increases per capita infrastructure-related expenditure in those regions. This result is consistent with the estimation results of previous studies showing regional bias in public capital investment in Japan.

[2] If there is no ruling party member in the area, the variable is missing. However, the conclusion does not change significantly even if the case where there is no ruling party member is treated as a dummy variable.

Most local public capital investments in Japan are based on the central government's transfer of financial resources. The ruling party's parliamentarians play a major role in transferring financial resources from the central government to public capital investment. Our estimation results in column (4) in Table 2.2 also support these effects. We calculated the magnitudes of the effect of the ruling party in the same manner as previously described. One standard deviation in the ratio of ruling party parliamentarians increases the ratio of infrastructure-related expenditure by 0.036. Comparing the magnitudes of coefficients while considering the standard deviation, the effect of ruling party parliamentarians is 11.7% (=0.036/0.311) of that of aging. Therefore, the effect of the 65-year-old and older population dominates that of ruling party parliamentarians on infrastructure-related expenditure. The first-stage estimation using the instrumental variable method is shown in Table 2.2. The results in the first stage show that the population ratio of those aged 65 and older for the past 10 years has a positive correlation with the current population ratio of those aged 65 years and older. The F value in the first stage had a large value. This suggests that the instruments were not weak.

Discussion on the efficiency of public capital investment

As mentioned in Sect. 2.2, several studies in Japan have analyzed the productivity and efficiency of public capital. The estimation results of Iwamoto (1990) and Mitsui and Ohta (1995) suggest that the accumulation of social capital stock in the 1980s was under the optimal size in Japan. Yoshino and Nakano (1994) show that the marginal productivity of public capital in urban areas is higher than that in rural areas. The accumulation of public capital stock in is low in urban areas and high in rural areas. Similar estimation results were shown by Asako et al. (1994), Doi (1998), and Yamano and Ohkawara (2000). Hayashi (2009) shows that the marginal productivity of public capital in the 2000s is on a recovery trend compared to before. If the public capital stock in Japan is excessive at the present time, the decrease in public capital investment due to aging will lead to the public capital stock coming closer to the optimal level. In addition, if the public capital stock is excessive in rural areas of Japan, aging will have a mitigating effect on excess capital stock in rural areas. Based on the instrumental variable method in Table 2.2, the estimation result shows that the higher the ratio of members of the ruling party, the more public capital investment is carried out. The ruling party in Japan is the LDP. The ratio of its members is high in rural areas. Local capital investments might not make much progress in mitigating excessive capital stock in rural areas because of the political process of members of the ruling party in these regions. Our estimation results show that the effect of aging on public capital investments dominates that of ruling party parliamentarians. Therefore, it is possible that aging brings public capital stock close to the optimal level in Japan, as these are excessive in Japan. Aging can derive efficiency in Japan in the short term. However, as aging progresses, even in urban areas, it negatively impacts efficiency in the long term.[3]

Discussion regarding the change of government

The estimation in Table 2.2 includes data from 1996 to 2014, during which period the Democratic Party of Japan (DPJ) was the ruling party between 2009 and 2012. The LDP had held power almost uninterrupted until then. The LDP was formed in 1955 through a merger between the Liberal Party and the Democratic Party. The LDP was in the opposition only twice: (i) from 1993 to 1994, when a coalition was formed by parties that were neither the LDP nor the Japanese Communist Party, and (ii) from 2009 to 2012, when the DPJ took control of the government. Except for the periods from 1993 to 1994 and from 2009 to 2012, the LDP has continuously been in power, sometimes ruling by itself or by forming a coalition. The so-called *zoku giin* (tribal lawmakers) or *doro zoku* (road-tribe lawmakers) stayed in the ruling party to influence the budget allocation process for public capital investments or infrastructure-related expenditures. Specifically, they promoted public capital investments in regional districts. Regarding the estimation of instrumental variables in Table 2.2, it has been statistically confirmed that the higher the ratio of ruling-party parliamentarians in a specific district, the higher the district's per capita infrastructure-related expenditures. Next, did the change in government influence the correlation between the ratio of ruling party parliamentarians and per capita infrastructure-related expenditures? The data in this study cover the period during which the DPJ controlled the government from 2009 to 2012. An analysis is conducted for this period by creating dummy variables named "democratic". It took 1 for 2009 to 2012 and 0 for other periods. The impact of the DPJ's control of the government is examined by adding "democratic × lratio ruling." It is a cross-term for "democratic" and "lratio ruling," which is the variable for the ratio of ruling-party parliamentarians. Table 2.3 only shows the fixed effects and instrumental variable estimation results. It confirms that the ratio of people aged 65 or above negatively impacts per capita infrastructure-related expenditure in the instrumental variable method. As for lratio ruling, it is not statistically significant, even though the coefficient is positive in the fixed-effects model in Table 2.2. On the other hand, in the fixed-effects model in Table 2.3, the coefficient of lratio ruling is positive and statistically significant. Meanwhile, in the instrumental variable method, the coefficient is positive and statistically significant in both Tables 2.2 and 2.3. Regarding the coefficients for "democratic," a dummy variable for the DPJ rule in Table 2.3, it is negative and statistically significant both in the fixed-effects model and the instrumental variable method. The coefficient for democratic × lratio ruling, the cross-term for the DPJ administration dummy ("democratic") and the ratio of ruling-party parliamentarians are negative for both the fixed-effects model and the instrumental variable method. However, this difference was not statistically significant. It is believed that the DPJ administration

[3] It is important to investigate whether public capital stock in Japan is excessive or insufficient at the macro level and whether it is excessive or insufficient in rural areas using datasets in recent years. This issue is left for future research.

Table 2.3 Estimation results: considering change of government

	Fixed Effect	IV	IV First stage
	(1)	(2)	lratio_aged65
lratio_aged65	0.1749	−0.6866*	
	(0.2724)	(0.3790)	
lratio_aged15	−0.0232	0.3720	0.6833***
	(0.5162)	(0.5333)	(0.1312)
lincome	1.8313***	1.3570**	−0.2986*
	(0.5064)	(0.6012)	(0.1358)
lgdp	−0.4786	−0.1562	0.2601*
	(0.5830)	(0.6795)	(0.1429)
lratio_ruling	0.0482*	0.0623**	0.0071
	(0.0243)	(0.0257)	(0.0056)
election	−0.0041	−0.0043	−0.0011
	(0.0081)	(0.0080)	(0.0022)
election_L1	−0.0173*	−0.0187**	−0.0031*
	(0.0086)	(0.0089)	(0.0016)
election_F1	0.0131*	0.0134*	−0.0009
	(0.0072)	(0.0070)	(0.0017)
lfiscalpower	−0.0918	−0.1062	−0.0068
	(0.0705)	(0.0741)	(0.0165)
lactive_open	0.1714***	0.1635***	0.0098*
	(0.0298)	(0.0323)	(0.0056)
trend	−0.0454***	−0.0186	0.0052*
	(0.0111)	(0.0133)	(0.0029)
year2008	−0.0546***	−0.0521***	0.0000
	(0.0151)	(0.0147)	(0.0014)
democratic	−0.0980***	−0.1100***	−0.0232***
	(0.0242)	(0.0231)	(0.0037)
lratio_ruling ×	−0.0752	−0.0786	0.0039
democratic	(0.0513)	(0.0517)	(0.0078)
lratio_aged65			0.8850***
_10yearsago			(0.1078)
sigma_u	0.4900	0.5539	0.0788
sigma_e	0.1209	0.1244	0.0259
rho	0.9425	0.9519	0.9019
R-squared			

(continued)

Table 2.3 (continued)

	Fixed Effect	IV	IV First stage
	(1)	(2)	lratio_aged65
Within	0.8551	0.8465	0.9716
Between	0.0658	0.3446	0.7729
Overall	0.1555	0.0458	0.8671
Observations	868	868	868

Note *p < 0.1; **p < 0.05; ***p < 0.01. Cluster-robust standard errors are in parentheses. Column (1) shows an estimation result based on the fixed-effect model, and column (2) shows estimation results based on the instrumental variable method

pursued a policy of allocating more resources for education and welfare rather than for public capital investments under the slogan "from buildings to people." The estimation results in the instrumental variable method demonstrate that the coefficients for "democratic" are negative and significant, confirming a nationwide decline in the level of infrastructure-related expenditures. Regarding the impact of the ratio of ruling party parliamentarians, the coefficient of democratic × lratio ruling is also negative in the instrumental variable method. However, it was not statistically significant. If this coefficient were significant, it could be statistically confirmed that the higher the ratio of ruling party parliamentarians in a specific district under the DPJ administration, the lower the district's per capita infrastructure-related expenditures. However, this effect was not statistically significant.

2.6 Concluding Remarks

Using Japanese prefectural data, we investigated whether aging reduces public capital investment. Estimation results based on the instrumental variable method show that aging reduces public capital investment, even considering the influence of local government infrastructure spending on the residential mobility of older people. Further, results show that Japanese public capital investments have a regional bias due to political processes. The acquisition of a budget for investment in public capital is considered as this process. The ruling party's parliamentarians played a major role in it. Our estimation results show that the higher the ratio of local parliamentarians, the higher the proportion of infrastructure-related spending. This suggests that the political process works in the bias of local public capital investment. In Japan, aging in rural areas is more advanced than in urban areas. The ratio of members of the ruling party's parliamentarians in the region is also higher in rural areas. This partially offsets the decrease in public capital investments in rural areas. However, the effect of aging in reducing investments in public capital is larger and dominates that of the ruling party's parliamentarians in the region. If capital stocks are excessive in rural areas of Japan, aging can resolve them in the short term. Conversely, public capital investment in urban areas is considered insufficient, and aging is expected to enhance it. Aging can bring about an efficient level in the short term because of the mitigation of excess capital stock in rural areas, but it is expected to enhance the insufficiency of capital stock in urban areas when aging progresses there. Thus, aging has a negative impact on efficiency in the long term.

The LDP has ruled Japan almost uninterruptedly. However, as mentioned, regime change occurred twice: from 1993 to 1994 and from 2009 to 2012. During the period analyzed in this study, the DPJ, rather than the LDP, was the ruling party from 2009 to 2012. It was confirmed that per capita infrastructure-related expenditures declined nationwide under the DPJ administration. However, no change has been observed in the relationship between the ratio of ruling party parliamentarians in regional areas and per capita infrastructure-related expenditures. Based on this, it may be inferred that, when it comes to making decisions about public capital investments in Japan, the politics underlying national treasury disbursements involving regionally elected parliamentarians did not disappear even after the regime change.

References

Arrow KJ, Kurz M (1970) Public investment, the rate of return, and optimal fiscal policy. Johns Hopkins University Press, Baltimore

Asako K, Tsuneki A, Fukuda S, Teruyama H, Tsukamoto T, Sugiura M (1994) Productivity of public capital and welfare analysis on public capital investment (*Syakai shihon no seisan-ryoku kōka to kōkyo-tōshi seisaku no kōsei-hyōka*). Econ Anal, No. 135 (in Japanese)

Aschauer DA (1989) Is public expenditure productive? J Monet Econ 23:177–200

Bacchiocchi E, Borghi E, Missale A (2011) Public investment under fiscal constraints. Fisc Study 32:11–42

Doi T (1998) Panel analysis of public capital in Japan (Nihon no syakai-shihon ni kansuru paneru bunseki). Kokumin Keizai 161:27–52 (in Japanese)

Harris AR, Evans WN, Schwab RM (2001) Education spending in an aging America. J Public Econ 81:449–472

Hayashi M (2009) Productivity effects of public capital: Reconsideration based on dynamic panel (Kōkyo-shihon no seisan-kōka: Dōgaku paneru ni yoru sai-kō). Zaisei Kenkyu 5:119–140 (in Japanese)

Iwamoto Y (1990) An evaluation of public investment policy in postwar Japan (Nihon no kōkyo-tōshi seisaku no hyō-ka ni tsu i te). Econ Rev 41:250–261 (in Japanese)

Jäger P, Schmidt T (2016) The political economy of public investment when population is aging: a panel cointegration analysis. Eur J Polit Econ 43:145–158

Kondoh H (2008) Political economy on public capital formation in Japan (Syakai-shihon seibi ni o ke ru seiji-keizai teki sokumen). Financial Rev 89:68–92 (in Japanese)

Ladd HF, Murray SE (2001) Intergenerational conflict reconsidered: county demographic structure and the demand for public education. Econ Educ Rev 20(4):343–357

Mehrotra A, Välilä T (2006) Public investment in Europe: evolution and determinants in perspective. Fisc Study 27:443–471

Mitsui K, Ohta K (1995) Productivity of public capital and government financial institution (Syakai-shino no seisan-sei to kōteki kinyu). Nippon Hyoron Sya, Tokyo (in Japanese)

Nordhaus WD (1975) The political business cycle. Rev Econ Stud 42:169–190

Vuchelen J, Caekelbergh S (2010) Explaining public investment in Western Europe. Appl Econ 42:1783–1796

Yamano N, Ohkawara T (2000) The regional allocation of public investment: Efficiency or equity? J Reg Sci 40:205–229

Yoshino N, Nakano H (1994) Regional distribution of public investment and productivity effects (Kokyo-tōshi no chiiki haibun to seisan-kōka). Financial Rev 41 (in Japanese)

Chapter 3
Effects of Aging on Education and Welfare Expenditures in Japan

3.1 Introduction

Investment in education is a public expenditure that provides long-term benefits, like infrastructure investment. The impact of population aging on education spending theoretically has both positive and negative effects. Elderly people past the schooling age are likely to support increased spending that will directly benefit the elderly, instead of increased education spending. Population aging can therefore be expected to have a negative impact on education spending. However, it is also possible that the elderly might support increased education spending. It is assumed that young people's productivity increases because of increased education spending. As the wages of the working-age population rise because of increased productivity, the elderly would also benefit from the increased expenditure transfer from the social security system, such as pensions and healthcare. In such cases, the elderly would likely favor increasing education spending. Furthermore, the elderly would also likely support increased education spending if local property values increased. If enhanced public education improves standards of living or reduces crime rates, this will improve the living conditions in the area, increasing land prices, and benefiting elderly property owners and homeowners. Finally, the elderly might support increased spending on education for altruistic reasons. As higher education spending increases the productivity of their children and grandchildren, it increases utility because of increased consumption by their children and grandchildren, which would increase the utility of altruistic elderly people. This chapter analyzes the impact of population aging on education spending using Japanese prefectural data.

Unlike infrastructure-related and education investments, spending on welfare has no long-term benefits. Thus, as the population ages, the elderly might likely favor increased expenditures on social welfare and elderly welfare, leading to an increase in these expenditures. We use Japanese prefectural data to analyze the impact of aging on expenditure on social welfare and elderly welfare, besides education expenditures. Furthermore, we empirically analyze public assistance expenditures which are

K. Terai et al., *The Political Economy of Population Aging*, Advances in Japanese Business and Economics 30, https://doi.org/10.1007/978-981-16-5536-4_3

unlikely to offer long-term benefits, such as social welfare and welfare expenditures for the elderly.

To an extent, local governments can set budgets for education and social welfare expenditure. Particularly, owing to the introduction of long-term care insurance in 2000, local governments' discretion to provide welfare services for the elderly has increased. We focus on the discretion of local governments. Furthermore, before 2000, elderly nursing homes provided most elderly welfare services. The admission capacity of these homes was important for providing elderly welfare services. Our analysis will also focus on the effect of admission capacity on welfare expenditures for the elderly.

The structure of this chapter is as follows. Section 3.2 reviews previous research related to education and welfare expenditure. Section 3.3 provides an overview of the features of Japan's education and welfare systems. Section 3.4 details the data and estimation methods used in this study. Section 3.5 details the estimation results, and Sect. 3.6 presents the concluding remarks.

3.2 Literature

Poterba (1997) empirically analyzed the impact of aging on government spending, focusing on how aging affects education spending using U.S. state data from 1960 to 1990. This study found that aging negatively impacted education spending. Harris et al. (2001) also analyzed the impact of aging on education spending using U.S. district data. Their estimation results confirmed the negative impact of aging on education spending, but the results were more modest than those in previous studies. Krieger and Ruhose (2013) used data from OECD countries to analyze the effect of aging on government spending. They used data from 22 OECD countries from 1985 to 2005 and analyzed whether aging affects childcare-related spending and education spending. According to the results of the panel analysis, aging does not have a significant negative impact on these expenditures. Sanz and Velázquez (2007) also conducted a panel analysis using data from OECD countries to analyze the effects of aging on social welfare spending. The results showed that aging increases social welfare and health care expenditures and, consequently, increases government expenditure. The study also found that aging was correlated with increased defense expenditures. Studies have empirically analyzed how aging affects public capital investment.

Ladd and Murray (2001) analyzed the effect of aging on public educational spending using state-level panel data in the United States from 1970 to 1990. They conducted an instrumental variable method to consider the endogeneity problem for elderly residents. Their estimation results showed that aging has no impact on educational spending, and they stated that one cannot predict the impact of an increasing share of the elderly on education spending without considering how elderly people would be distributed. Figlio and Fletcher (2012) also considered an endogeneity problem for the share of the elderly in a community. They matched the 1950 and

1960 population and housing censuses with recent data to deal with endogeneity problems. Their estimation results showed that the elderly percentage in a school district was negatively related to the level of support for public schooling. Regarding European countries, Grob and Wolter (2007) used Swiss Cantons data, and Borge and Rattso (2008) used Denmark's local government-level data to estimate the effect of aging on educational spending. Both estimation results show that the elderly population share has a negative impact on spending on public education.

Ohtake and Sano (2009) and Miyaki and Kimura (2019) analyzed the impact of population aging on education expenditure using Japanese prefectural data. Ohtake and Sano (2009) show that population aging had a positive effect on compulsory education expenditure before 1985, and it had a negative impact after 1990. In contrast, Miyaki and Kimura (2019) show that, while in the first half of the 2000s, the elderly who supported high education spending levels on high school and university-level education, have not been supportive of increased expenditure on higher education levels since the second half of the 2000s.

Some analyses of government spending in Japan, particularly local government spending, are concerned with the effects of municipal mergers. Nakazawa (2013) quantitatively analyzes certified processes for long-term care and insurance premiums in municipalities that have undergone municipal mergers and shows that municipalities experiencing unsound financial situations tend to have more generous certification for long-term care and low long-term care insurance premiums. Hirota and Yunoue (2008) focus on nursing insurance finances and show that operating through *Kouiki-rengo* or wider area union (increasing the number of people insured in a municipality) results in reduced annual expenditures. Miyazaki (2013) and others use Japanese data to analyze the effects of mergers on long-term care and overall local government spending. Reingewertz (2012) uses Swiss data to analyze the effects of municipal mergers on spending reductions. Abe et al. (2008) and Hayashi (2019) analyzed public assistance and local governments. Abe et al. (2008) discuss the impact of central government guarantees on public assistance to local governments. Hayashi (2019) uses Japanese data to analyze the impact of central government assistance on local governments' decisions regarding the certification of public assistance.

3.3 An Overview of Education and Welfare Systems

The Japanese schooling system is based on the School Education Law, which defines "schools" as kindergartens, elementary schools, junior high schools, high schools, special needs schools, and universities, where elementary and junior high schools are compulsory. Additionally, the duration of enrollment should be six years for elementary school, three years for junior high school, three years for high school, and four years for university. The central government sets the standards for the establishment of elementary, junior high, and high schools, alongside universities, standards for school curricula, and national standards for the number of teaching staff. It also covers one-third of teaching staff salaries in prefectural schools and one-half of overhead costs, such as school buildings. Local governments, enact ordinances for the number of teaching staff at prefectural schools, salaries, working conditions, appointments, dismissals, and disciplinary measures. Local governments also bear the cost of teaching staff salaries in public elementary, junior high, and high schools. In 2017, the top items in prefectural expenditure by purpose were: elementary school expenses, 27.8%; junior high school expenses, 16.6%; high school expenses, 20.8%; and education administration expenses, 20.8%. The top items in municipal educational expenses by purpose were elementary schools (27.6%), junior high schools (15.9%), social education expenses (15.4%),[1] health and physical education (19.0%), and education administration expenses (14.3%).[2]

Japan's welfare policy is implemented through social welfare and public assistance systems. The social welfare system is primarily concerned with the welfare of the elderly, the disabled, and children. Concerning elderly welfare, the focus is on long-term care support for the elderly, although there is a difference in the scope of the system before and after 2000. Before 2000, there was spending on projects such as the development and operation of long-term care facilities for the elderly in need of nursing care. When the long-term care insurance system was established in 2000, many policies for those aged over 65 and in need of nursing care were brought within the social insurance framework. Consequently, most of the expenditure on long-term care is done through the social insurance system, but long-term care for those aged 65 and above with poor economic or family circumstances is still provided through elderly welfare.

Welfare for the disabled is for those with physical, intellectual, and mental disabilities. According to the Ministry of Health, Labour and Welfare, in 2014, the total number of persons with disabilities was 7,789,000, of which 3,397,000 had physical disabilities, 741,000 had intellectual disabilities, and 3,201,000 had mental disabilities. Moreover, 69% of persons with physical disabilities, 9% of persons with intellectual disabilities, and 36% of persons with mental disabilities were over the age of 65. Hence, 50% of the total number of persons with disabilities were elderly people aged 65 and above. The top three items in the total costs of disability welfare services are long-term care for daily life (37.6%), support for continuous employment (15.3%), and support for institutionalization (11.1%).

Child welfare is for children and families with children. One specific example is the preschool. Recently, dual-income households have become common in Japan, and there has been an increase in the number of households using preschools. Earlier, single-parent families with working mothers and who could not stay home and look after the children were the focus of policies for households facing difficulties in their economic and familial circumstances. Consequently, the welfare system continues

[1] Expenses for social education facilities established under local government ordinances within the jurisdiction of the board of education and expenses for social education activities conducted by the board of education (including athletics, cultural activities, and protection of cultural properties).

[2] Ministry of Internal Affairs and Communications, "2019 White Paper on Local Government Finances (*Heisei 31-nendo-ban chihō zaisei hakusho*)".

to handle childcare policies. Child welfare other than preschools includes placement in homes for infants, children's homes, and children's self-reliance support facilities. These facilities are for children who require care in a home environment, such as those without guardians or victims of child abuse, and have upper age limits of 15 or 18 years. Other foster parent systems are also included in the child welfare system.

Public assistance is established by the central government but is operated by local governments that establish welfare offices. Each local government calculates the minimum cost of living for a household based on the household composition, age, and place of residence of the welfare applicant according to the standards determined by the central government (standard of livelihood protection). If households cannot make ends meet even after utilizing their income, savings, and assets, the shortfall is paid as welfare. There are broadly two standards for the minimum cost of living set by the central government. The first is the age and size of the household, and the second is the place of residence. The minimum cost of living varies by age and by the number of persons in the household. It also varies depending on where the applicant lives, such as whether they live in urban or suburban areas. In this system, the decision to provide living support payments to welfare applicants is made by the local governments, but the minimum cost of living standards, which forms the basis of welfare payments, is determined by the central government. According to the Ministry of Health, Labour and Welfare, the percentage of welfare recipients aged 65 and above was 47.4% in 2016.[3]

3.4 Data and Estimation Strategy

We used the prefectural basic data in the "Social and Demographic System" provided by the Ministry of Internal Affairs and Communications. This dataset is long-term panel data, and it is possible to conduct a panel analysis that considers individual effects in regions. The dataset contains general account budgets for each local government, total production, income, fiscal power index, and population by age in each prefecture. The basic formula for the estimation is as follows:

$$y_{i,t} = \alpha + \beta x_{i,t} + \beta^j X^j_{i,t} + \gamma_i + u_{i,t} \tag{3.1}$$

Here, the explained variables are per-capita educational expenditure, per-capita social welfare expenditure, per-capita elderly welfare expenditure, and per-capita public assistance expenditure in each prefecture. The data in general account expenditure provides total educational expenditure, total social welfare expenditure, total elderly welfare expenditure, and total public assistance expenditure. We calculated per-capita variables by dividing these items by the total population. Additionally,

[3] Ministry of Health, Labour and Welfare's document from the investigative commission on the development of new methods to verify public assistance standards, "Summary of the public assistance system (*Seikatsu hogo seido no gaiyō-tō ni tsuite*)".

we take the logarithm of these per-capita variables and use them as the explained variables. γ_i is an individual effect, and $u_{i,t}$ is an error term. Aging is the variable of interest in this study. We divide the population aged 65 and older as the total population in each prefecture and use its logarithm as a variable of aging (lratio_aged65). $X^j_{i,t}$ are the control variables. We use the logarithm of the ratio of the population under 15 years old (lratio_aged15), the logarithm of per-capita gross production (lgdp), the logarithm of per-capita income (lincome), and the logarithm of fiscal power index (lfiscalpower) in each prefecture. We include the log of the active opening ratio (lactiveopen) to control for the job market and conditions in the region. We include the log of the active opening ratio (lactiveopen) to control for the job market and conditions in the region. We divide the nominal value by the deflator in each prefecture to calculate the real terms in the GDP and per-capita income. We include per-capita gross production to control for yearly economic fluctuations in each region. Additionally, we include yearly trend terms to control for long-term socioeconomic trends in each region. From a political economy's viewpoint, politicians will take action to set budgets to gain an advantage in their election. Governor elections are important at the local government-level. Therefore, we elected the dummy variable (election) that takes the value of 1 when the governor's election is held, and otherwise 0 in each prefecture. We include this variable in the estimation to consider the effects of the governor's election on the budget of the local government. Additionally, there may have already been operating budgets before the governor's election-year. Therefore, we make a lead variable of the governor's election-year (election_F1) and add it to the estimation. Furthermore, we add a lag variable of the governor's election-year (election_L1) in the estimation. This variable is considered a reward for supporting the candidate after the governor's election. We calculate the ratio of the ruling party members to the elected total parliamentarians in each prefecture and take a log (lratio_ruling). We include this variable in the estimations. We calculate the ratio of the ruling party members to the elected total parliamentarians in each prefecture and take a log (lratio_ruling). We include this variable in the estimations in the same way as in Chap. 2.

We conducted a panel analysis that considers individual effects in regions in estimation (3.1). However, if local governments' budgets affect the population of those aged 65 and older in the region, the estimated coefficient of ratio_65 has a bias. People aged 65 and older may migrate to municipalities that provide higher values for budget items preferred by elderly people. Then again, they may leave local governments where the budget preferred by the elderly is low. If there are residential movements of elderly people between local governments, the estimation in formula (3.1) has a bias. We used the instrumental variable method to deal with the problem of elderly residential movements driven by budget allocation in local governments. Assuming that the budget value by item affects the population ratio for people aged 65 and older, the explanatory variable correlates with the error term, and consistency in the estimation cannot be guaranteed, even with the fixed-effect model or the random-effect model, in the panel analysis. In this case, the instrumental variable method can be used to obtain a consistent estimator.

Table 3.1 Descriptive statistics

	Mean	Std. Dev.	Min	Max
Dependent variables				
education	93.911	15.452	58.048	164.669
Elderlywelfare	13.532	7.295	2.124	36.679
Socialwelfare	10.027	5.141	1.755	28.582
Publicassistance	2.711	2.068	0.199	13.756
Explanatory variables				
ratio_aged65	0.189	0.053	0.071	0.326
ratio_aged15	0.158	0.028	0.108	0.273
income	2563.164	419.264	1665.222	4421.053
gdp	3.274	0.714	1.862	7.189
activeopen	1.695	0.840	0.409	6.413
fiscalpower	0.474	0.221	0.197	1.639
election	0.255	0.436	0	1
election_L1	0.253	0.434	0	1
election_F1	0.263	0.440	0	1
ratio_rulling	0.640	0.246	0	1
trend	25.500	8.658	11	40
year2008	0.033	0.179	0	1
Instrument				
ratio_aged65	0.141	0.044	0.052	0.267
_10yearsago				
Observations	1410			

$$x_{i,t} = \tau + \pi z_{i,t} + \epsilon_{i,t} \tag{3.2}$$

Here, z is an instrumental variable, and it is required that there be no correlation with the error term $u_{i,t}$ in (3.1). In this chapter, we use the log of the ratio of people aged 65 and above to the total population for the past 10 years (lratio_aged65past) in the same way as in Chap. 2. The ratio of people aged 65 and above in the past has been correlated with the current aged population ratio (lratio_aged65). Conversely, it is not affected by the current per-capita education and welfare expenditures. Therefore, the variable lratio_aged65past is considered an appropriate instrumental variable. Table 3.1 shows the descriptive statistics of variables in the estimations.

3.5 Estimation Results

Table 3.2 shows the results of a simple OLS estimation, which does not consider regional individual effects. The sample period is from 1985 to 2014. The results

Table 3.2 Estimation results in pooled OLS

	Education	Socialwelfare	Elderlywelfare	Publicassitance
	(1)	(2)	(3)	(4)
lratio_aged65	0.0252	0.3329	0.5804***	−0.0568
	(0.0589)	(0.3110)	(0.2061)	(0.6018)
lratio_aged15	0.2679***	0.0714	−0.0734	−0.2945
	(0.0857)	(0.4573)	(0.3133)	(0.9163)
lincome	0.3181***	−0.2420	−0.2676	−0.6149
	(0.0560)	(0.4641)	(0.3042)	(0.9974)
lgdp	0.3337***	0.4229	0.7752**	0.2850
	(0.0596)	(0.5144)	(0.3107)	(0.6941)
lratio_rulling	−0.0745***	−0.1254	0.0061	0.1075
	(0.0191)	(0.0657)	(0.0623)	(0.1269)
election	−0.0049	−0.0205*	−0.0052	0.0270
	(0.0033)	(0.0115)	(0.0073)	(0.0231)
election_L1	−0.0006	−0.0149*	−0.0111	0.0258
	(0.0032)	(0.0086)	(0.0074)	(0.0202)
election_F1	−0.0038	−0.0027	−0.0093	0.0367*
	(0.0031)	(0.0110)	(0.0068)	(0.0202)
lfiscalpower	−0.4784***	−0.2974***	−0.3898***	−0.9548***
	(0.0295)	(0.0830)	(0.0514)	(0.2828)
lactiveopen	0.0424***	−0.0533**	0.0744***	0.3896***
	(0.0094)	(0.0261)	(0.0221)	(0.0974)
trend	0.0002	0.0364***	0.0316***	−0.0281***
	(0.0015)	(0.0077)	(0.0038)	(0.0087)
year2008	−0.0012	0.1245***	0.1299***	−0.0943
	(0.0067)	(0.0238)	(0.0125)	(0.0651)
R-squared	0.8074	0.7183	0.9006	0.5075
Observations	1410	1410	1410	1410

Note *$p < 0.1$; **$p < 0.05$; ***$p < 0.01$. Cluster-robust standard errors are shown in parentheses

show that the ratio of the population aged 65 and above had a positive sign on per-capita expenditures on education, social welfare, and elderly welfare, but a negative sign on per-capita public assistance expenditure. The coefficient of the aging variable in elderly welfare is only statistically significant. Table 3.3 presents the estimation results using a fixed-effect model, considering regional individual effects. Column (1) in Table 3.3 shows the estimates of per-capita education expenditure as the dependent variable. The results show that the coefficient for the ratio of the population aged 65 and above is positive, but it is not statistically significant. Similarly, the coefficient for the ratio of the population under the age of 15 was positive but statistically

Table 3.3 Estimation results in fixed-effect model

	Education	Socialwelfare	Elderlywelfare	Publicassitance
	(1)	(2)	(3)	(4)
lratio_aged65	0.0656	0.8282***	0.3404*	2.0561***
	(0.0723)	(0.2898)	(0.2015)	(0.5157)
lratio_aged15	−0.1893	0.7983**	−0.1719	0.8664
	(0.1331)	(0.3509)	(0.2761)	(0.5644)
lincome	0.4457***	1.3667***	0.7305**	0.0485
	(0.1082)	(0.3248)	(0.3510)	(0.6817)
lgdp	0.2612*	−1.0983***	−0.6604*	−0.0693
	(0.1462)	(0.3388)	(0.3651)	(0.6805)
lratio_rulling	−0.0151	0.0320	0.0306	0.1580*
	(0.0124)	(0.0412)	(0.0398)	(0.0914)
election	−0.0025	−0.0203**	−0.0068	0.0058
	(0.0017)	(0.0086)	(0.0078)	(0.0103)
election_L1	0.0009	−0.0085	−0.0072	0.0034
	(0.0021)	(0.0063)	(0.0072)	(0.0114)
election_F1	−0.0025	−0.0103	−0.0122	0.0075
	(0.0019)	(0.0093)	(0.0074)	(0.0087)
lfiscalpower	−0.1006***	1.0443***	−0.0175	−0.8633***
	(0.0320)	(0.1069)	(0.0750)	(0.1644)
lactiveopen	0.0436***	−0.0094	0.0892***	0.1102***
	(0.0073)	(0.0220)	(0.0158)	(0.0360)
trend	−0.0085***	0.0426***	0.0454***	−0.0681***
	(0.0024)	(0.0064)	(0.0058)	(0.0173)
year2008	−0.0114**	0.0073	0.0599***	0.0136
	(0.0048)	(0.0186)	(0.0094)	(0.0236)
sigma_u	0.1796	0.5721	0.2454	0.5700
sigma_e	0.0462	0.1716	0.1296	0.2437
rho	0.9379	0.9174	0.7818	0.8454
R-squared				
Within	0.6062	0.8652	0.9389	0.4583
Between	0.0537	0.4678	0.0674	0.3943
Overall	0.0024	0.1658	0.7702	0.3986
Observations	1410	1410	1410	1410

Note *p < 0.1; **p < 0.05; ***p < 0.01. Cluster-robust standard errors are shown in parentheses

insignificant. The coefficient for per-capita income within the prefecture—the variable for regional income—is positive and statistically significant. Per-capita GDP within the prefecture, indicating economic fluctuations in regions, has a positive impact on education. We also include governor election year dummies and dummies for the years before and after governor elections as election-related variables; neither yielded statistically significant results.

Next, we investigate the impact of population aging on per-capita social welfare expenditure and per-capita elderly welfare expenditure. Column (2) of Table 3.3 shows the estimation result by applying a fixed-effect model for per-capita social welfare expenditure. The coefficient for the ratio of the population aged 65 and above was positive and statistically significant. Additionally, the coefficient for the ratio of the population aged under 15 years is positive and statistically significant. A social welfare system includes expenditures for young people, such as child welfare. This has a positive effect. The coefficient of per-capita income within the prefecture is positive and statistically significant. This can be interpreted to mean that local government tax revenues are high in regions where incomes are high, therefore the local government can offer more generous social welfare expenditure. The coefficient of per-capita GDP by prefecture is negative and statistically significant. The coefficient of per-capita GDP by prefecture is negative and statistically significant. This can be interpreted that per-capita social welfare expenditures fell in regions where economic conditions are better off and per-capita social expenditures increased in regions where economic conditions are worse off. Hence, we can infer that social welfare expenditure has a built-in stabilizing effect. The coefficient of the governor election year dummy is negative and statistically significant. Owing to a built-in stabilizer effect, there is a possibility that social welfare expenditures decrease if economic conditions are good in the governor election year. Column (3) in Table 3.3 shows the estimated result taking per-capita expenditure on elderly welfare as the explained variable. The coefficient for the ratio of the population aged 65 and above was positive and significant. Conversely, the coefficient for the ratio of the population aged under 15 years is negative but not statistically significant. Regarding the governor election year dummy and dummies for the years before and after the elections, these coefficients are not statistically significant.

Next, we investigate the effects of population aging on per-capita public assistance expenditures. The results are shown in Column (4) of Table 3.3. The coefficient for the ratio of the population aged 65 and above was positive and statistically significant. This means that as the elderly population increases, so does per-capita public assistance expenditure. The prefectural per-capita GDP coefficient was not statistically significant. Public assistance is also expected to provide a built-in stabilizer effect; however, the signs of the coefficients and their lack of statistical significance prevent us from confirming these effects. Neither the governor election year dummy nor the dummies for the years before and after elections are statistically significant.

Instrumental variable method

For per-capita educational expenditure, the estimates using the instrumental variable method are shown in Column (1) of Table 3.4. The coefficient for the ratio of the

Table 3.4 Estimation results in IV

	Education	Socialwelfare	Elderlywelfare	Publicassitance	First stage
	(1)	(2)	(3)	(4)	lratio_aged65
lratio_aged65	−0.6903***	1.3594**	0.8363	4.1338***	
	(0.2498)	(0.6525)	(0.5505)	(0.9160)	
lratio_agcd15	−0.5308***	1.0382**	0.0521	1.8051***	−0.2870***
	(0.1703)	(0.4509)	(0.3858)	(0.6956)	(0.1048)
lincome	0.3134***	1.4596***	0.8173**	0.4121	−0.0575
	(0.1435)	(0.3613)	(0.4024)	(0.6693)	(0.0968)
lgdp	0.3286*	−1.1457***	−0.7047*	−0.2547	0.1822*
	(0.1689)	(0.3646)	(0.3923)	(0.7312)	(0.1086)
lratio_rulling	0.0012	0.0204	0.0198	0.1129	0.0102
	(0.0146)	(0.0475)	(0.0397)	(0.0923)	(0.0095)
election	−0.0026	−0.0202**	−0.0068	0.0060	−0.0015
	(0.0021)	(0.0085)	(0.0078)	(0.0112)	(0.0015)
election_L1	0.0012	−0.0087	−0.0075	0.0025	−0.0016
	(0.0024)	(0.0062)	(0.0071)	(0.0122)	(0.0014)
election_F1	−0.0011	−0.0112	−0.0132	0.0037	0.0007
	(0.0020)	(0.0093)	(0.0070)	(0.0090)	(0.0013)
lfiscalpower	−0.1891***	1.1064***	0.0405	−0.6203***	−0.1448***
	(0.0509)	(0.1319)	(0.0778)	(0.2257)	(0.0381)
lactiveopen	0.0377***	−0.0052	0.0931***	0.1266***	0.0103*
	(0.0078)	(0.0219)	(0.0167)	(0.0364)	(0.0054)
trend	0.0079	0.0310**	0.0345***	−0.1134***	0.0053
	(0.0052)	(0.0138)	(0.0114)	(0.0248)	(0.0033)
year2008	−0.0169***	0.0112	0.0635***	0.0286	−0.0145***
	(0.0058)	(0.0181)	(0.0115)	(0.0279)	(0.0034)
lratio_aged65					0.5899***
_10yearsago					(0.1108)
sigma_u	0.2265	0.5447	0.2176	0.7075	0.0280
sigma_e	0.0576	0.1733	0.1315	0.2615	0.0403
rho	0.9391	0.9080	0.7322	0.8797	0.3248
R-squared					
Within	0.3860	0.8625	0.9370	0.3761	0.9780
Between	0.3641	0.4358	0.2864	0.3138	0.9654
Overall	0.0928	0.1955	0.8073	0.3226	0.9749
Observations	1410	1410	1410	940	1410

Note *$p < 0.1$; **$p < 0.05$; ***$p < 0.01$. Cluster-robust standard errors are shown in parentheses

population aged 65 and above was negative and statistically significant. The impact of population aging on education spending could, in theory, be positive or negative, or cancel out. Japanese data from 1995 to 2014 show negative impacts. For the other variables, the ratio of the population under 15 was negative and statistically significant under the instrumental variable method. The governor election year dummy and dummies for the year before and after elections are likewise not statistically significant. The ratio of the population aged 65 and above for the past 10 years in the first stage is positive and statistically significant. Moreover, the F-value in the first stage was also high and significant.

Column (2) in Table 3.4 shows the estimation result taking per-capita social welfare expenditure as the explained variable, while Column (3) in Table 3.4 shows the estimation of per-capital expenditure on elderly welfare as the explained variable. Endogeneity problems are expected to arise concerning social welfare and welfare expenditures for the elderly. We also use the ratio of the population aged 65 and above 10 years prior as the instrumental variable in both cases. The estimation of per-capital social welfare expenditure (Column (2) of Table 3.4), as with the fixed-effect model, shows that the coefficient for the ratio of the population aged 65 and above is positive and statistically significant. Moreover, the coefficient for prefectural per-capita GDP is negative and statistically significant which implies a built-in stabilizing effect. The coefficient for the fiscal power index, showing the local governments' financial status, is positive and statistically significant. From this, it can be interpreted that municipalities with greater financial strength are more generous with social welfare expenditure. Using the instrumental variable method for per-capital expenditure on elderly welfare (Column (3) of Table 3.4), the coefficient for the ratio of the population aged 65 and above is positive but not statistically significant. As for elderly welfare expenditure, the effect of aging on this expenditure is not statistically confirmed when we employ the instrumental variable method. Governor election dummies were not significant. Concerning the explanatory variables, we use the same variables as those used in the education expenditure analysis and the instrumental variable. The first stage is the same as in the case of educational expenditure; the ratio of the population aged 65 and above 10 years prior is positive and significant, and the F-value is also high.

Column (4) of Table 3.4 shows the results for per-capita public assistance expenditure. The results are statistically significant for the coefficient of the ratio of the population aged 65 and above. Based on the positive sign, aging leads to an increase in per-capita public assistance expenditure. Moreover, the coefficient for the ratio of the population aged under 15 years is also positive and statistically significant. There is a framework within public assistance whereby additional benefits are paid for children; thus, these additional benefits can be interpreted as having a positive effect. The governor election year dummy and dummies for the year before and after are not statistically significant. The first stage is the same as that for education, social welfare, and welfare expenditures for the elderly.

Impact of the introduction of long-term care insurance

The expenditures on elderly welfare may have been influenced by the introduction of long-term care insurance in 2000. As stated earlier, before 2000, welfare for the elderly focused on the provision of services through the elderly welfare system. Since 2000, with the introduction of long-term care insurance, the services provided by the elderly welfare system, such as admission to special nursing homes, are now provided under a separate account—long-term care insurance—and the majority of the existing elderly welfare services were transferred to long-term care insurance. Considering the effects of this change in the system, we performed an analysis after narrowing the sample to the post-2000 period. The results are presented in Table 3.5. The impact of the ratio of the population aged 65 and above on per-capita expenditure on elderly welfare is positive and statistically significant in the period following the introduction of long-term care insurance. By transferring the previous elderly welfare system to the social insurance framework, it became possible for a person of any income level to receive benefits and services as long as they pay their insurance premium contributions. Besides, those insured under long-term care insurance are based on local governments, and additional services can now be provided through independent policies by local governments. It can, therefore, be considered that since 2000, welfare services for the elderly provided through long-term care insurance have become easier to expand at the local governments' discretion. The elderly welfare system is for those to whom services cannot be provided under the long-term care insurance system because of their economic or familial situation, but these services overlap with those of long-term care insurance in many respects. The expansion of welfare services under the long-term care insurance system can, therefore, be expected to be accompanied by an expansion in services under the elderly welfare system. One reason for the significance of the variable for the ratio of population aged 65 and above after 2000 could be because of an increased discretion of local governments to provide welfare services for the elderly through long-term care insurance and the accompanying expansion of the elderly welfare system.

As for other variables, the coefficient for the ratio of the population aged under 15 was found to be negative and significant in Table 3.5. Since the coefficient for the ratio of population aged under 15 is negative, a higher population aged under 15 would result in a reduction in per-capita expenditure on elderly welfare. These results suggest an intergenerational conflict in expenditure on welfare for the elderly.

Discussion on the differences between the systems

We discuss the institutional impact based on an estimation using the instrumental variable method. While it depends on the sample period, we find that the ratio of the population aged 65 and above had a negative impact on per-capita education expenditure, a positive impact on per-capita social welfare expenditure, per-capita elderly welfare expenditure, and public assistance expenditure. First, the difference between education expenditure and the impact on social welfare and public assistance expenditures can be interpreted as a difference in the benefits of these expenditures on the elderly population. Concerning education expenditure, if there is little effect in terms

Table 3.5 Estimation results in IV: considering the introduction of long-term care insurance

	Elderlywelfare	First stage
		lratio_aged65
lratio_aged65	0.7679*	
	(0.4341)	
lratio_aged15	−1.1433***	0.4685***
	(0.3916)	(0.1236)
lincome	0.8524***	−0.1041
	(0.2533)	(0.0874)
lgdp	−0.4089	0.1290
	(0.3256)	(0.0899)
lratio_rulling	−0.0413	0.0200***
	(0.0358)	(0.0071)
election	−0.0121*	−0.0025
	(0.0069)	(0.0024)
election_L1	−0.0038	−0.0030
	(0.0043)	(0.0018)
election_F1	−0.0071	0.0007
	(0.0066)	(0.0020)
lfiscalpower	−0.0825	0.0324*
	(0.0509)	(0.0189)
lactiveopen	0.2772***	−0.0004
	(0.0147)	(0.0044)
trend	0.0325***	0.0042
	(0.0097)	(0.0034)
year2008	−0.0100	−0.0131***
	(0.0109)	(0.0016)
lratio_aged65		0.7691***
_10yearsago		(0.1185)
sigma_u	0.1738	0.0591
sigma_e	0.0819	0.0220
rho	0.8183	0.8785
R-squared		
Within	0.9083	0.9643
Between	0.3598	0.7778
Overall	0.6832	0.8608
Observations	705	705

Note *p < 0.1; ***p < 0.01. Cluster-robust standard errors are shown in parentheses

of a systematic income transfer to the elderly, such as public pension, because of higher wages for the current working population, then the benefits will not be felt by the elderly population. In such cases, they would not favor increased education expenditure. Conversely, the benefits of social welfare and public assistance expenditures are felt directly by the elderly population, and the elderly population will be in favor of an increase in expenditures. We believe that the estimation results show a difference in the benefits to the elderly population.

Concerning welfare expenditure, the elderly receive a direct benefit from these. However, the estimation result for the entire sample period using the instrumental variable method does not show this effect. As for the elderly welfare system before 2000, most of these services were provided in elderly nursing homes. If the admission capacity of elderly nursing homes is full, it is difficult for local governments to increase these services. Therefore, local governments could not increase these services before 2000, even if they wanted to do so. Conversely, regarding the public assistance system, the level of protection for welfare recipients is set uniformly by the central government. Thus, local governments cannot independently increase the benefit level of public assistance. However, local governments can decide to provide living support payments to welfare applicants. They can control the number of welfare applicants and increase those numbers at the request of local residents because there is no admission capacity for public assistance in principle. Thus, the difference in the impact on public assistance and elderly welfare expenditures before 2000 can be interpreted as the difference in the capacities of these services.

3.6 Concluding Remarks

In this study, we analyze the impact of population aging on education expenditure, social welfare expenditure, elderly welfare expenditure, and public assistance expenditure. Welfare expenditures, such as social welfare, elderly welfare, and public assistance expenditures, provide direct benefits to the elderly, but education expenditure does not. Estimation results using the instrumental variable method confirm that an increase in the proportion of the population aged 65 and above leads to an increase in per-capita social welfare expenditure and per-capita public assistance expenditures, and reduces per-capita expenditure on education. These results suggest that an increase in population aging leads to increased expenditure that directly provides benefits to the elderly, while tending to reduce expenditures that offer long-term benefits, such as education expenditure, not providing benefits to the elderly. Conversely, it was confirmed that aging led to an increase in per-capita elderly welfare expenditure after the introduction of long-term care insurance in 2000. Since 2000, welfare services for the elderly provided through long-term care insurance have become easier to expand at the local governments' discretion. Therefore, the increased local governments' discretion to provide welfare services for the elderly through long-term care insurance is considered the result of increasing per-capita elderly welfare expenditure after 2000. Additionally, most services of the elderly welfare system were provided by elderly nursing homes. The local governments had difficulty increasing these services when admission capacities in nursing homes were full. The difference in the impact on public assistance and elderly welfare expenditures before 2000 can be interpreted as the difference in the capacities of these services.

References

Abe A, Kunieda S, Suzuki W, Hayashi M (2008) Economic analysis of public assistance (Seikatsu hogo no keizai bunseki). University of Tokyo Press, Tokyo (in Japanese)

Borge LE, Rattso J (2008) Young and old competing for public welfare service. CESifo Working Paper 2223

Figlio DN, Fletcher D (2012) Suburbanization, demographic change and the consequences for school finance. J Public Econ 96(11–12):1144–1153

Grob U, Wolter SC (2007) Demographic change and public education spending: a conflict between young and old? Educ Econ 15(3):277–292

Harris AR, Evans WN, Schwab RM (2001) Education spending in an aging America. J Public Econ 81(3):449–472

Hayashi M (2019) Do central government grants affect welfare caseloads? Evidence from public assistance in Japan. Finanzarchiv 75(2):152–186

Hirota H, Yunoue H (2008) Does border-based local government affect expenditure on public long-term care insurance? The case of Japan. Econ Bull 8(11):1–20

Krieger T, Ruhose J (2013) Honey, I shrunk the kids' benefits—revisiting intergenerational conflict in OECD countries. Public Choice 157(1–2):115–143

Ladd HF, Murray SE (2001) Intergenerational conflict reconsidered: county demographic structure and the demand for public education. Econ Educ Rev 20(4):343–357

Miyaki M, Kimura M (2019) Population aging and shifts in public education expenditure: What levels of education do the elderly support? (Jinkō kōrei-ka to kō kyōiku-hi no hensen - kōrei-sha wa dono kyōiku dankai o shiji suru ka). Jpn Econ Res 77:61–88 (in Japanese)

Miyazaki T (2013) Cost reduction and economic of scale: Evidence from municipal consolidation in Japan. Working Paper, Kyusyu University

Nakazawa K (2013) Free-ride or adjustment? Amalgamation and long-term care insurance premium setting in Japanese municipality. Working Paper, Toyo University

Ohtake F, Sano S (2009) Population aging and compulsory education expenditure (Jinkō Kōrei-Ka to Gimukyōiku Shishutsu). Osaka Univ Econ 59(3):106–130 (in Japanese)

Poterba JM (1997) Demographic structure and the political economy of public education. J Policy Anal Manage 16(1):48–66

Reingewertz Y (2012) Do municipal amalgamation work? Evidence from municipalities in Israel. J Urban Econ 72(2–3):240–251

Sanz I, Velázquez FJ (2007) The role of aging in the growth of government and social welfare spending in the OECD. Eur J Polit Econ 23(4):917–931

Chapter 4
Effects of Aging on Corporate Taxes in Japan

4.1 Introduction

This chapter focuses on analyzing the effects of aging on corporate tax rates. As seen in Chaps. 2 and 3, population aging increases Japan's social welfare spending per capita through political processes; on the other hand, it decreases the per capita spending on other areas that result in long-term benefits, such as infrastructure and education in Japan. Subsequently, another question is how aging affects corporate activities that support local economic growth and employment. In this chapter, we analyze those problems focusing on statutory corporate tax rates. Empirical works such Kneller et al. (1999), Lee and Gordon (2005), and Arnold (2008) indicate that the negative effect of corporate taxes on economic growth is relatively larger than that of other taxes. If aging leads to higher corporate tax rates, it slows economic growth through the channel of corporate tax rates. Meanwhile, many elderly citizens own assets such as real estate and stocks. Therefore, if lower corporate tax rates stimulate corporate activities and lead to increases in the value of the real estate and stocks they own, they may not oppose corporate tax rate reductions.

Figure 4.1 shows the changes in the statutory tax rate of corporate income tax in Japan. The statutory tax rate, which combines corporate taxation and local corporate taxation, kept declining through the 1950s and 1960s. In the 1970s, it was gradually increased until it reached its peak in 1984. Afterwards, government adopted a worldwide trend of low corporate taxes and reduced its rates. In the 2000s, major reductions began to be implemented. The statutory tax rate of local corporate taxation (Corporate Inhabitant Tax + Corporate Enterprise Tax) followed a similar trend, although to a lesser degree than that of the national corporate tax. In 1955, it was 15.53% on a standard tax rate basis, but it gradually decreased and reached 14.86% in 1964. Subsequently, it started to increase again until 1984, when it peaked at 16.94%. Since 1987, it switched to a downward trend again but was still considered relatively high. Hence, it went through significant cuts in 1998 and 1999 to be closer to international standards. The corporate tax system in Japan is detailed later, but since corporate taxes are determined uniformly by the government for the entire

K. Terai et al., *The Political Economy of Population Aging*, Advances in Japanese Business and Economics 30, https://doi.org/10.1007/978-981-16-5536-4_4

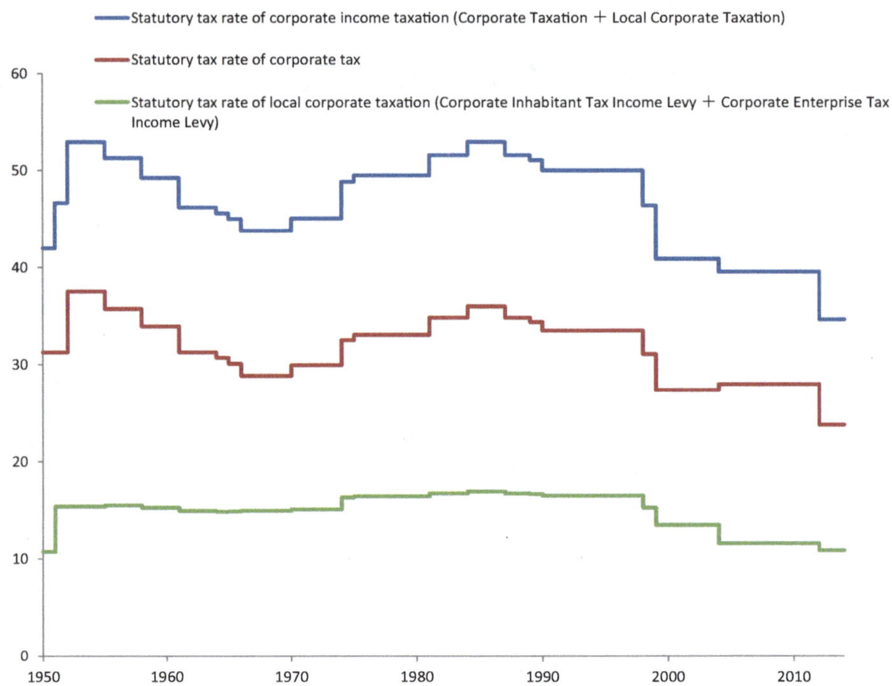

Fig. 4.1 Changes in the statutory tax rate of corporate income taxation in Japan Standard tax rate basis. *Source* Ministry of Internal Affairs and Communications "Local Corporation Taxation Related Materials"

country, they do not vary from region to region. Conversely, local corporate taxation may have different rates depending on the region because of practices such as tax exemption and non-uniform taxation stipulated by local governments. In this chapter, we conduct an empirical analysis of local corporate taxation using those variations among different regions.

The structure of this chapter is as follows. Section 4.2 reviews previous research on corporate taxes. Section 4.3 provides an overview of the features of Japan's corporate tax system. Section 4.4 details the data and estimation method used in this chapter. Section 4.5 details the estimation results and, finally, Sect. 4.6 states the conclusion.

4.2 Literature

As mentioned in the introduction, previous studies have conducted empirical analyses of the relationship between corporate taxes and economic growth. Kneller et al. (1999) empirically analyzed how the composition of taxes and government spending affects economic growth based on the Barro-type growth regression model. They used panel data from 22 OECD countries from 1970 to 1995. They found that the heavier the "distortionary" income taxes, including corporate, the lower the economic growth. Lee and Gordon (2005) analyzed the effects of corporate taxes on economic growth using data from more than 70 countries, including developing countries, from 1970 to 1997. Based on their estimates, they claim that a corporate tax reduction of 10 percentage points increases the economic growth rate from 1 to 2%. Arnold (2008) analyzed the taxation and economic growth from 1971 to 2004 using panel data from 21 OECD countries. Based on the estimation results, it concluded that a fixed asset tax has the smallest impact on economic growth, followed by a consumption tax. Conversely, it stated that a corporate tax has the most negative impact.

Meanwhile, there are a few empirical studies in the context of tax competition among neighboring municipalities. Chirinko and Wilson (2008) empirically analyzed corporate tax competition using data from U.S. states. Instead of supporting tax competition, their findings indicate that the fact that a certain state is influenced by the corporate tax rates around it reflects the same response to common shocks between that state and surrounding states. Hayashi and Boadway (2001) conducted a panel analysis using data from Canadian provinces. In their estimates, they used the average tax rate, which is calculated by dividing the tax revenue by the corporate income, as an explained variable, instead of the statutory tax rate. Based on the results, they concluded that the province of Ontario is affected by the tax rates of other provinces, showing that tax competition exists. Buettner (2001) used a smaller administrative division and focused on cities, instead of states, to analyze tax competition in Germany. To this end, they used instrumental variables. The results show that, at the city level, the business taxes of adjacent municipalities affect each other, which supports the idea of tax competition. Further, cities with smaller populations tend to set lower tax rates. Leprince et al. (2007) conducted an analysis by the instrumental variable method using data from France. Their empirical results confirmed that there is business tax competition among French regions.

In Japan, the taxes on corporate income include Corporate Tax, Local Corporate Tax, Local Inhabitant Tax, Local Enterprise Tax, and Local Corporate Special Tax. Fukasawa (2009) focused on the Local Enterprise Tax, which follows relatively easy rules set by each local government, to analyze whether there was tax competition on a prefectural basis. He used panel data from 46 prefectures in Japan (only Tokyo Metropolis was excluded) from 1986 to 1993 and 1997 to 2005. The data from 1997 confirmed that the average Corporate Enterprise Tax rate of each prefecture was reduced following the reduction in the average tax rate in neighboring prefectures, and the estimation results support the existence of tax competition. Conversely, this effect cannot be clearly identified in the data prior to 1993.

4.3 Overview of Corporate Tax Systems in Japan

There are five main corporate taxes in Japan: (1) Corporate Tax (national), (2) Local Corporate Tax (national), (3) Special Corporate Enterprise Tax (national),

(4) Corporate Inhabitant Tax (prefectural/municipal), and (5) Corporate Enterprise Tax (prefectural). Local Corporate Tax is a part of Corporate Inhabitant Tax (income levy) converted into a national tax. It is designed to correct the uneven distribution of tax sources between regions and reduce the financial disparity between them. It was created in 2014 as a national tax to secure a source of tax revenue to be allocated to local governments following the reduction in the Corporate Inhabitant Tax rate. Additionally, the Special Corporate Enterprise Tax was introduced in October 2019; before that, there was a system called Local Corporate Special Tax, which was also a national tax.

Figure 4.2 shows an overview of corporate taxation. Tax rates are based on 2014 values. First, the amount of Corporate Tax, a national tax, is calculated by multiplying the corporate income by the tax rate. Subsequently, this amount of Corporate Tax is multiplied by the Local Corporate Tax rate to derive the Local Corporate Tax. The Corporate Inhabitant Tax, a local tax, is charged as income levy and per capita basis. Income levy is calculated by multiplying the amount of Corporate Tax by the Local Inhabitant Tax rate. Income levy of the Local Corporate Tax is charged as prefectural and municipal taxes. Also, the per capita basis is collected at a fixed amount that varies for each municipality. The Corporate Enterprise Tax is calculated by multiplying the corporate income by the Corporate Enterprise Tax rate.

In 2004, there was the introduction of pro forma standard taxation (size-based business tax). In the case of ordinary corporations with a capital of over 100 million yen, half of their corporate income is subject to Corporate Enterprise Tax, and the remaining half is subject to pro forma standard taxation, which consists of a value-added levy tax and per capita tax. The former is calculated by multiplying the value-added (revenue distribution amount + single year profit/loss) by the tax rate, and the latter, by multiplying the amount of capital by the tax rate. Ordinary corporations and public-service corporations with less than 100 million yen in capital are not subject

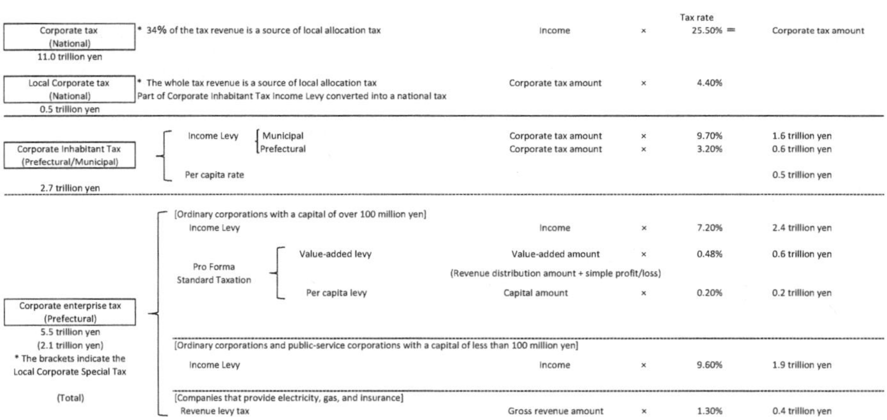

Fig. 4.2 Overview of corporate taxation in Japan. *Source* Ministry of Internal Affairs and Communications "Local Corporation Taxation Related Materials". *Note* Tax rates and revenues are based on 2014 values

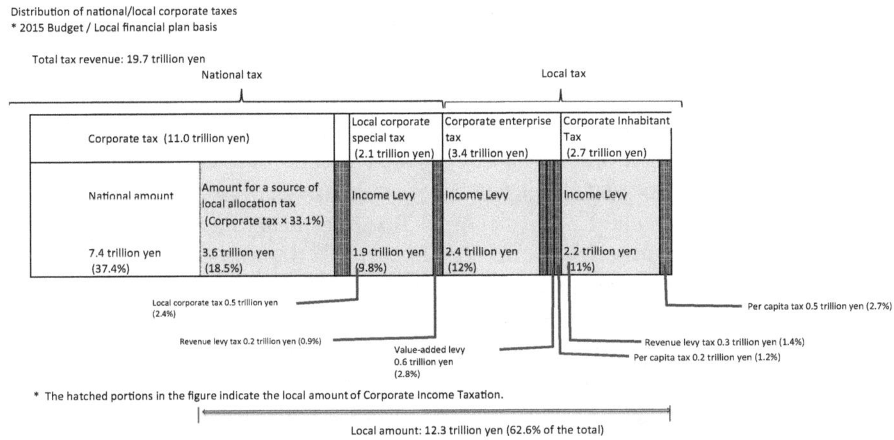

Fig. 4.3 Distribution of national/local corporate taxes in Japan. *Source* Ministry of Internal Affairs and Communications "Local Corporation Taxation Related Materials"

to pro forma standard taxation; their Corporate Enterprise Tax amount is calculated by multiplying their corporate income by the Corporate Enterprise Tax rate. Also, for companies that provide electricity, gas, and insurance, the taxes are calculated by multiplying their income by the tax rate.

One of the characteristics of corporate taxes in Japan, as mentioned in the introduction, is that Tokyo's corporate tax revenue is outstandingly high. To correct this geographic unevenness, corporate tax revenue is redistributed to other regions as a source of local allocation tax. Figure 4.3 shows the national and local allocation of corporate taxes. The total corporate tax revenue in 2015, combining national and local taxes, was 19.7 trillion yen, but national taxes (Corporate Tax, Local Corporate Tax, and Local Corporate Special Tax) account for about 70% of the total with 13.6 trillion yen. Local taxes, in the form of Corporate Enterprise Tax and Corporate Inhabitant Tax, are responsible for about 30% of the total. Meanwhile, one-third of the corporate tax revenue becomes a source of local allocation tax to be distributed to other regions and correct geographical disparities. Additionally, the Local Corporate Tax was originally the Local Inhabitant Tax, and the Local Corporate Special Tax was originally the Corporate Enterprise Tax. Local Corporate Tax and Local Corporate Special Tax are first collected as national taxes and, after the correction of the uneven distribution of tax sources, they become local financial resources. After these corrections, a total of 12.3 billion yen, which is more than 60% of the total, becomes local financial resources. These indicate that corporate taxation also partially contributes to tax redistribution among regions.

4.4 Data and Estimation Strategy

We use the prefectural basic data in the "Social and Demographic System" provided by the Ministry of Internal Affairs and Communications, as in Chaps. 2 and 3, to conduct a panel analysis. Companies in Japan are basically subject to Corporate Tax, Corporate Inhabitant Tax, and Corporate Enterprise Tax. The Corporate Tax includes National Corporate Tax and Local Corporate Tax, but the tax rate is uniform across the country. The Corporate Inhabitant Tax is calculated on a per capita basis and income levy, and their values and rates are slightly different for each municipality. Further, many municipalities have their own rules regarding Corporate Enterprise Tax and exempt some companies from it or apply non-uniform taxation to stimulate the regional economy. For this reason, the Corporate Enterprise Tax rate also varies from municipality to municipality. However, since there are no data that systematically summarize the differences in tax rates among those municipalities, for this study, we calculate the average statutory tax rate for each prefecture by dividing the corporate tax revenue of each prefecture by the corporate income. The corporate income of each prefecture can be obtained from the "Social and Demographic System", and the corporate tax revenue, from the "Survey of Local Public Finance" (Ministry of Internal Affairs and Communications). The population structure, local GDP, and income of each prefecture were also obtained from the "Social and Demographic System". The basic formula for estimation is as follows:

$$y_{i,t} = \alpha + \beta x_{i,t} + \beta^j X^j_{i,t} + \delta \sum\nolimits_{j \neq i} \omega_{i,j} w_{i,t-1} + \gamma_i + u_{i,t} \tag{4.1}$$

Here, the explained variable is the average statutory corporate tax rate in each prefecture. We calculate it by dividing the revenue of corporate tax items by corporate incomes. γ_i is an individual effect, and $u_{i,t}$ is an error term. Aging is a variable of interest in this chapter like in Chaps. 2 and 3. We divide the population aged 65 and older by the total population in each prefecture and use the logarithm of it as a variable of aging (lratio_aged65). $X^j_{i,t}$ are control variables. We use the logarithm of total population (lpopulation), the logarithm of per capita gross production (lgdp), the logarithm of income per capita (lincome), and the logarithm of fiscal power index (lfiscalpower) in each prefecture. We include the log of the active opening ratio (lactiveopen) to control for the job market and conditions in the region. We divide the nominal value by the deflator in each prefecture to calculate real terms in capita gross domestic product and income per capita. We include total population to control for the effect of agglomeration and per capita gross production to control yearly economic fluctuation in each region. In addition, we include the yearly trend term to control for long-term socio-economic trends in each region. From a political economy's point of view, we include the governor's election. We make the election a dummy variable (election) in each prefecture that takes a value of 1 when the governor's election was held or 0 otherwise. Having a member of parliament from the ruling party may make it easier to promote policies designed to attract businesses.

Therefore, we include the ratio of the number of the members of the ruling party to the total members of parliament members in each prefecture.

Presumably, the definition of corporate tax rates by each local government is influenced by the policies of other municipalities around them. For this reason, we include the corporate tax rate of surrounding municipalities as an explanatory variable. $w_{i,t-1}$ is the average statutory tax rate, calculated by dividing the corporate tax revenue of each prefecture by the corporate income. $\omega_{i,j}$ adds the weight of neighboring prefectures to this average statutory tax rate. δ is the coefficient of the effect of the average corporate statutory tax rate on the concerned municipality.

As the tax rates of the concerned prefecture and the prefectures around it affect each other, the problem of endogeneity occurs. To counter this problem of endogeneity, the tax rates of the surrounding prefectures in the previous period were added as explanatory variables, similar to Clark and Lohéac (2007). The tax rate of the concerned prefecture does not affect the surrounding tax rates of the previous period that have already been determined. Hence, the surrounding tax rates of the previous period were used as explanatory variables. The weights of $\omega_{i,j}$ were calculated with three weights. The first method (Weight 1) is based on simple adjacency between municipalities; that is, the weight takes 1 if it shares a prefectural border with the prefecture in question and 0 otherwise. The next method (Weight 2) uses the inverse distance to each prefecture as the weight, rather than adjacency. The denominator of the weight is the total sum of the inverse distance to each prefecture, and the numerator is the inverse distance to each prefecture. The last method (Weight 3) is similar to Weight 2; it multiplies the inverse distance to each prefecture used in Weight 2 by the population of each prefecture.

This chapter also makes the estimation using the instrumental variable method. As welfare spending, discussed in Chap. 3, brings direct benefits to the elderly, regions with high levels of welfare spending may have a strong influence on the elderlies' decision to move. Even though corporate taxes do not impact the elderly directly, i f policies such as corporate tax exemption lead to more jobs for the young people in the corresponding region, it may affect its ratio of elderly population. To take this effect into account, this chapter also uses the instrumental variable method:

$$x_{i,t} = \tau + \pi z_{i,t} + \epsilon_{i,t} \tag{4.2}$$

Here, z is an instrumental variable, and it is required that it has no correlation with the error term $u_{i,t}$ in (4.1). We use the logarithm of the ratio of people aged 65 and above to the total population for the past 10 years (lratio_aged65past) in the same way as in Chaps. 2 and 3. The ratio of people aged 65 and above in the past is correlated with the current aged population ratio (lratio_aged65). Conversely, it is not affected by the current corporate statutory tax rate (lcorporate_tax). Therefore, the variable of lratio_aged65past is considered to be an appropriate instrumental variable. Table 4.1 shows the descriptive statistics of variables in the estimations.

Table 4.1 Descriptive statistics

	Mean	Std. Dev.	Min	Max
Dependent variables				
corporate_tax	37.737	15.369	11.732	106.332
Explanatory variables				
ratio_aged65	0.221	0.039	0.110	0.326
population	2711194	2577478	577000	13400000
income	2585.198	399.294	1818.347	4418.649
gdp	3.409	0.665	2.365	7.189
activeopen	1.886	0.802	0.516	6.413
election	0.256	0.436	0	1
ratio_rulling	0.699	0.264	0	1
neighbor_tax				
contiguity	37.935	12.572	14.598	76.498
distance	38.705	11.208	20.438	65.085
distance and population	44.813	12.778	24.043	78.787
trend	31.500	5.191	23	40
year2008	0.055	0.229	0	1
Instrument				
ratio_aged65	0.166	0.038	0.075	0.267
_10yearsago				
Observations	846			

4.5 Estimation Results

Tables 4.2 and 4.3 show the estimation results of the ordinary least squares and the fixed-effect model, respectively. In Sect. 4.5, we mention the results of the fixed-effect model that controls for individual effects of the regions. Column (1) of Table 4.3 shows the results estimated in τ using the average statutory tax rate of neighboring prefectures of the previous period. Those results indicate that the sign of the ratio of the population aged 65 and above is negative but not statistically significant. The variables of effective job openings-to-applicants ratio and income per capita are included to control for the local labor market and income level. Those coefficients are negative and statistically significant. The variable of average statutory tax rate takes the local corporate income as the denominator. If the local labor market conditions are bad, the corporate income also tends to decrease. Also, if the local income per capita is low, the local purchasing power is low, and the corporate income of such regions tends to be low. The negative coefficients indicate these relationships. The sign of the coefficient of the GDP per capita of the region is positive but not statistically significant.

Further, the region's percentage of ruling party members in the parliament was added as a political economy variable, but it is not statistically significant. The same

Table 4.2 Estimation results in pooled OLS

	Contiguity	Distance	Distance and population
	(1)	(2)	(3)
lratio_aged65	−0.7756**	−0.5465***	−0.5412***
	(0.3089)	(0.1844)	(0.1822)
lpopulation	0.1522***	0.1537***	0.1573
	(0.0548)	(0.0452)	(0.0461)
lactiveopen	−0.4458***	−0.4522***	−0.4453***
	(0.0396)	(0.0349)	(0.0346)
lincome	−1.9553***	−1.4252***	−1.5008***
	(0.3469)	(0.2808)	(0.2725)
lgdp	1.2028***	0.9956***	1.0997***
	(0.2570)	(0.2278)	(0.2203)
lratio_rulling	0.0796	0.0580	0.0638
	(0.0994)	(0.0978)	(0.0985)
election	0.0016	0.0057	0.0044
	(0.0088)	(0.0090)	(0.0092)
lneighbor_tax_t−1	0.4489***	0.6206***	0.6071***
	(0.0534)	(0.0356)	(0.0392)
trend	−0.0149*	−0.0126**	−0.0149***
	(0.0087)	(0.0055)	(0.0054)
year2008	−0.1221***	−0.0971***	−0.0982***
	(0.0215)	(0.0201)	(0.0205)
R-squared	0.7213	0.7415	0.7402
Observations	810	846	846

Note *p < 0.1; **p < 0.05; ***p < 0.01. Cluster-robust standard errors are in parentheses

is true for the governor election dummy. The average statutory corporate tax rate of neighboring prefectures of the previous period, δ, has a positive sign, and it is statistically significant. Column (2) of Table 4.3 is an estimation of the statutory corporate tax rate of neighboring prefectures in the previous period using a variable weighted by the inverse distance as δ. The coefficient of the ratio of the population aged 65 and above is negative but not significant. The results of the control variables are same as those in Column (1). Also, the coefficient of δ is positive and statistically significant.

Column (3) of Table 4.3 is an estimate of the statutory tax rate of neighboring prefectures in the previous period using a variable weighted by the inverse the distance to neighboring prefectures and population as δ. Like above, the coefficient of the ratio of the population aged 65 and above is negative but not significant. The same result was obtained for the other control variables. The coefficient of the statutory tax rate δ weighted by the distance and population is also positive and statistically

Table 4.3 Estimation results in fixed effect model

	Contiguity	Distance	Distance and population
	(1)	(2)	(3)
lratio_aged65	−0.2143	−0.3785	−0.3295
	(0.4173)	(0.3906)	(0.3965)
lpopulation	0.4283	0.8231	0.7551
	(0.5630)	(0.5161)	(0.5146)
lactiveopen	−0.4850***	−0.4795***	−0.4750***
	(0.0234)	(0.0219)	(0.0219)
lincome	−2.4028***	−2.1601***	−2.0749***
	(0.3449)	(0.3051)	(0.3027)
lgdp	0.7565	0.6040	0.5521
	(0.5108)	(0.4807)	(0.4787)
lratio_rulling	0.0599	0.0470	0.0497
	(0.0577)	(0.0545)	(0.0549)
election	0.0095	0.0090	0.0083
	(0.0072)	(0.0070)	(0.0071)
lneighbor_tax_t−1	0.4981***	0.5567***	0.5555***
	(0.0271)	(0.0259)	(0.0257)
trend	−0.0286**	−0.0204*	−0.0233*
	(0.0120)	(0.0115)	(0.0115)
year2008	−0.1150***	−0.1091***	−0.1086***
	(0.0151)	(0.0136)	(0.0136)
sigma_u	0.2181	0.4520	0.4051
sigma_e	0.1473	0.1426	0.1433
rho	0.6867	0.9094	0.8887
R-squared			
Within	0.8058	0.8138	0.8119
Between	0.3875	0.3593	0.3535
Overall	0.6164	0.4250	0.4477
Observations	810	846	846

Note *$p < 0.1$; **$p < 0.05$; ***$p < 0.01$. Cluster-robust standard errors are in parentheses

significant. These results indicate that the statutory corporate tax rate in Japan is affected by neighboring prefectures and even when it is weighted by the distance and the population, it is still heavily affected by the policies of neighboring prefectures.

Instrumental variable method

If a region has a high statutory corporate tax rate, companies may move from that region to an adjacent prefecture with a lower statutory tax rate, and the younger

Table 4.4 Estimation results in IV

	Contiguity	First stage	Distance	First stage	Distance and population	First stage
		lratio_aged65		lratio_aged65		lratio_aged65
	(1)		(2)		(3)	
lratio_aged65	−0.9543		−1.0265		−1.1017*	
	(0.7683)		(0.6506)		(0.6534)	
lpopulation	1.1877	0.7816***	1.3642**	0.6229***	1.3995***	0.6222***
	(0.7707)	(0.1840)	(0.5513)	(0.2025)	(0.5442)	(0.2022)
lactiveopen	−0.4931***	−0.0069	−0.4873***	−0.0076	−0.4843***	−0.0076
	(0.0235)	(0.0053)	(0.0219)	(0.0053)	(0.0219)	(0.0053)
lincome	−2.6642***	−0.1396	−2.3670***	−0.0986	−2.3233***	−0.1034
	(0.4294)	(0.1205)	(0.3559)	(0.1195)	(0.3572)	(0.1209)
lgdp	1.0453*	0.2164	0.7949	0.1353	0.7798	0.1362
	(0.5811)	(0.1134)	(0.5046)	(0.1249)	(0.5029)	(0.1254)
lratio_rulling	0.0816	0.0228***	0.0670	0.0234***	0.0734	0.0232***
	(0.0569)	(0.0083)	(0.0543)	(0.0077)	(0.0548)	(0.0078)
election	0.0080	−0.0018	0.0078	−0.0017	0.0068	−0.0018
	(0.0073)	(0.0013)	(0.00711)	(0.0012)	(0.0072)	(0.0012)
lneighbor_tax	0.5064***	0.0083	0.5647***	0.0077 *	0.5637***	0.0049
_t−1	(0.0264)	(0.0053)	(0.0252)	(0.0039)	(0.0251)	(0.0044)
trend	−0.0080	0.0077***	−0.0027	0.0054*	−0.0023	0.0053*
	(0.0214)	(0.0026)	(0.0179)	(0.0031)	(0.0178)	(0.0031)
year2008	−0.1184***	−0.0090***	−0.1122***	−0.0094***	−0.1125***	−0.0096***
	(0.0152)	(0.0021)	(0.0138)	(0.0020)	(0.0137)	(0.0020)
lratio_aged65		0.5895***		0.6416***		0.6425***
_10yearsago		(0.0859)		(0.0939)		(0.0936)
sigma_u	0.7518	0.6097	0.8896	0.4792	0.9230	0.4781
sigma_e	0.1487	0.0241	0.1439	0.0252	0.1451	0.0253
rho	0.9623	0.9984	0.9744	0.9972	0.9758	0.9972
R-squared						
Within	0.8020	0.9716	0.8104	0.9690	0.8071	0.9689
Between	0.4001	0.5257	0.3610	0.3205	0.3586	0.3197
Overall	0.3392	0.0755	0.2929	0.0159	0.2856	0.0157
Observations	810	810	846	846	846	846

Note *p < 0.1; **p < 0.05; ***p < 0.01. Cluster-robust standard errors are in parentheses

generations may follow suit. Ultimately, it affects the ratio of the local population aged 65 and above. To take the possibility of such endogeneity into account, we conduct an estimation using the ratio of the population aged 65 and above as of ten years before as an instrumental variable, as previously done in Chaps. 2 and 3. The estimation results are shown in Table 4.4.[1] Column (1) of Table 4.4 is the estimation result obtained using the average statutory corporate tax rate of neighboring prefectures in the previous period as δ.

The results are the same as those of the fixed-effect model in Column (1) of Table 4.3. The ratio of the population aged 65 and above does not become significant. Also, δ is positive and significant. The τ of Column (2) of Table 4.4 is the adjacent prefecture's statutory corporate tax rate in the previous period weighted by the inverse distance. Except for the fact that the total population of the corresponding region, lpopulation, became positive and significant, the results are the same as those in Column (1) of Table 4.4: the ratio of the population aged 65 and older is not significant. Additionally, the coefficient of δ is positive and statistically significant. Column (3) of Table 4.4 is the estimated statutory corporate tax rate of neighboring prefectures in the previous period using a variable weighted by the inverse distance to neighboring prefectures and the population of neighboring prefectures as δ. Based on this estimation model, the coefficient of the ratio of the population aged 65 and above was negative and statistically significant.

Presumably, many municipalities expect that by lowering corporate statutory tax rates, they can attract companies to their region and stimulate corporate activities. The fact that the coefficient of aging variable is negative and significant indicates that the elderly support the policy of stimulating local economic activities through corporate activities. Decreasing the corporate statutory tax rate of a municipality also decreases the tax resources of that municipality, but if it stimulates corporate activities and increases the local land prices, the elderly—who own a large percentage of assets—may agree with the decrease in the corporate statutory tax rate.

Further, in Japan, the gap between expenditure and revenue of local governments is filled with local allocation tax grants. This ensures that even if the corporate tax revenue of local governments decreases, the expenditures for the elderly are not cut immediately. In such circumstances, the link between reductions in corporate tax revenue and reductions in the spending for the elderly is not too strong, which suggests that the elderly may not strongly oppose reductions in the corporate statutory tax rate. The coefficient of δ is positive and statistically significant consistently. Regarding the effect of corporate statutory tax rates on neighboring prefectures, it remains positive even when they are weighted by the distance and the population.

4.6 Concluding Remarks

In this chapter, we analyzed whether aging affects corporate tax rates using prefectural-level data from Japan. Previous research has pointed out that corporate taxes in one prefecture are influenced by the taxes in its surrounding prefectures, and similar estimation results were obtained in this chapter. In the analysis of the relationship between aging and corporate tax rates, when the tax rates of the neighboring prefectures were controlled, the variable of aging was not significant. Conversely,

[1] The results are the same even when we exclude Tokyo, where corporate tax revenue is prominent.

in the estimation weighted by the inverse distance to other prefectures and population, the coefficient of the variable of aging was negative and significant. If lower corporate tax rates result in lower spending for the elderly, the elderly may oppose the reduction in corporate tax rates.

However, in Japan, there is a mechanism that adjusts shortages of local general financial resources through local allocation tax grants. There is also another mechanism that redistributes a certain percentage of corporate tax and enterprise tax (which are collected as national taxes) as local government resources through local allocation tax grants. With these measures, even if local governments reduce the corporate tax rates, they can still maintain the spending for the elderly to some extent. Furthermore, if reductions in corporate tax stimulate local corporate activities and raise the value of the local real estate and companies that operate in the region, the elderly—who own relatively large volumes of real estate and financial assets—may not oppose those reductions. The estimation results of this chapter suggest that the elderly would not oppose reductions in corporate taxes.

References

Arnold A (2008) Do tax structures affect aggregate economic growth? Empirical evidence from a panel of OECD countries. OECD Economics Department Working Papers 643

Buettner T (2001) Local business taxation and competition for capital: the choice of the tax rate. Reg Sci Urban Econ 31:215–245

Chirinko B, Wilson DJ (2008) Tax competition among U.S. states: racing to the bottom or riding on a seesaw? Working Paper Series, Federal Reserve Bank of San Francisco

Clark AE, Lohéac Y (2007) It wasn't me, it was them! Social influence in risky behavior by adolescent. J Health Econ 26:763–784

Fukasawa E (2009) Tax competition over taxation of local corporations in Japan-Analysis of the current situation of corporate enterprise tax (Waga-kuni no chiho-hojin kazei wo meguru sozei-kyōsō: Hōjin-jigyō zei wo taisyo to shi ta genjō bunseki). Reference 703:55–75 (in Japanese)

Hayashi M, Boadway R (2001) An empirical analysis of intergovernmental tax interaction: the case of business income taxes in Canada. Can J Econ 34(2):48–53

Kneller R, Bleaney MF, Norman G (1999) Fiscal policy and growth: evidence from OECD countries. J Public Econ 74:171–190

Lee Y, Gordon RH (2005) Tax structure and economic growth. J Public Econ 89:1027–1043

Leprince T, Madi'es T, Paty S (2007) Business tax interactions among local governments: An empirical analysis of the French case. J Reg Sci 47(3):603–621

Chapter 5
Effects of the Elderly Population and of Political Factors in the US States

5.1 Introduction

The US has a relatively smaller elderly population than Japan: 15% are over 65 years old in the US in 2015, compared to 26% in Japan. But by 2050 the US proportion is expected to reach 22%.[1] If the aged voters do not want to pay taxes for the programs benefiting children or young families, these programs may have smaller budgets.

A column in the Los Angeles Times on January 16, 2000, by Dana Parsons, describes how opposition by elderly voters caused a tax proposal in Irvine, California, fail. The author continues[2]:

> Some no doubt philosophically opposed the tax or are so strapped financially that it represented a hardship. But that can't be everyone-I'm thinking of seniors who own their homes outright and would gladly pony up $8 a month (the cost of the parcel tax) for other community projects. The questions are: Why not for schools? Why not for the current generation's future?

The column entitled "$95 Too Taxing for Greatest Generation?" places a focus on a cohort who was born between early 1900s and 1920s and survived tough periods of the Depression and World War II. Is this anti-tax sentiment shared only by the generation who sacrificed their lives and paid taxes to make the great economy? Or do old voters likely oppose tax increase that will be spent on programs for children and young families? The purpose of this chapter, using data from the 2000s and 2010s, is to examine how a higher share of the elderly voters in a state is associated with public spending that is expected to benefit younger generations.

The elderly in the US differ from the general population in several ways. Among the elderly , only 17% are Black or Hispanic, compared to 32% among those younger than 65. Among the elderly 3.5% were non-citizens, which is half the rate among

[1] United Nations Department of Economic and Social Affairs, World Population Prospects: The 2015 Revision; and US Census Bureau, Population Estimates and Projections.

[2] https://www.latimes.com/archives/la-xpm-2000-jan-16-me-54657-story.html.

K. Terai et al., *The Political Economy of Population Aging*, Advances in Japanese Business and Economics 30, https://doi.org/10.1007/978-981-16-5536-4_5

the general population. Though among male elderly the college completion rate is essentially the same as among the general population over age 25, elderly women are far less likely to have completed college: 22% versus 32% in the younger population. That low education rate may reduce the political influence of the elderly. And this difference is particularly important because women constitute many more of the elderly: 79 males to 100 females among the elderly, compared to a ratio of 97 in the general population.[3] The elderly population shows a wide variation across the states, ranging from 11% in Utah to 20% in Florida in 2017.[4] We exploit that variation and control other demographic factors to explore how the elderly affect state policy.

5.2 Related Literature

The political economics of taxation and redistribution was pioneered by Romer (1975), Roberts (1977), and Meltzer and Richard (1981). Their fruitful approach considers the median voter model with two political candidates and voters differing in their incomes and so in their preferences for taxes and spending; the equilibrium has the tax rate preferred by the voter with median income (Meltzer and Richard 1981). Another approach considers probabilistic voting (Hettich and Winer 1988, 1999), where government aims to maximize expected support, and the probability that a voter supports a tax increases with the difference between the value of benefits received and the costliness of taxes paid. Such models predict a complex income tax, with multiple brackets targeting particular constituencies.

The preferences of the majority party in a legislature, and hence party control of a legislature, may affect public policy (Garand 1988; Alt and Lowry 1994; Rogers and Rogers 2000; Besley and Case 2003). For example, Democratic control of the House is associated with higher government spending (Alt and Lowry 1994; Rogers and Rogers 2000). Democratic control of a legislature is associated with significantly higher taxes and a redistribution of spending in favor of family assistance (Besley and Case 2003).

Opposition to welfare is often rooted in prejudice towards poor Blacks and is a form of race-coding where white Americans can express their prejudice towards Blacks without being explicitly racist (Gilens 1996). The author further uses an experimental manipulation to show how perception of poor whites, or rather white welfare mothers, has less of an effect on opposition towards welfare than perception of poor Black welfare mothers. Soss et al. (2008) argue in their racial classification model that when minorities become more represented in the welfare discourse, race and racial classifications often assume salience to policy makers in their implementation of policies. The effect of immigration on welfare policies was explored by Hawes and McCrea (2018) who showed how immigration levels affect a community's social capital, which in turn affects its welfare generosity.

---

[3] https://www.census.gov/content/dam/Census/library/publications/2018/acs/ACS-38.pdf.

[4] These are estimates of the American Community Survey, US Census Bureau.

5.3 Empirical Analysis

To examine the effects of demographics, we use a panel data set of 49 states in the US between 2005 and 2017. The sample period is determined by data availability. We exclude Nebraska from our samples because the legislature of Nebraska is unicameral. The total number of observations is 637.

5.3.1 Data

We conduct empirical analysis for spending on selected policies that we regard as influenced by the elderly. Our regressions are

$$y_{it} = \alpha_i + x_{it}^{'}\beta + z_{it}^{'}\gamma + \varepsilon_{it}, \qquad (5.1)$$

where y_{it} denotes the share of state i's governmental spending on a particular policy in year t; α_i denotes an individual effect for state i; x_{it} is a vector of time-varying demographic variables that can influence the choice of spending by government in state i in year t; z_{it} is a vector of political or economic variables;β and γ are vectors of parameters to be estimated; ε_{it} is an error term for state i in year t.

The dependent variables are the percentage of total spending on education, highways, and public welfare. We use a proportion of total spending in a certain category, not its amount, because we are interested in the government's allocation of a given amount of revenue. Data for the dependent variables come from the Annual Survey of State and Local Government Finances, US Census Bureau.

The demographic variables included in x_{it} are percent population aged 65 and over, percent population under 15 years old, percent Black and percent Hispanic or Latino. The data are from the American Community Survey, US Census Bureau. We focus on the estimated coefficient of the percentage of the population aged 65 and over, which is expected to be negative when explaining the spending ratio on education and highways; otherwise, a positive coefficient is expected. The ratio of the budget allocated to investment in human and physical capital would decline with a larger share of the elderly population. If the estimated coefficient exceeds 1, it means that the share of total spending, e.g., on education, increases by over 1 percentage point with an increase of the share of the population that is elderly by 1 percentage point.

Political variables in z_{it} describe politicians' partisanship: a ratio of Democratic legislators in the state legislature, and a variable that takes 1 for Democratic governors and 0 otherwise.[5] If a governor is replaced by another governor in midyear, a weighted sum of these values is calculated on the basis of the length of incumbency.

[5] The data on governors are obtained from the National Governors Association. Also, the data on state legislators are collected by the National Conference of State Legislatures.

We included dummy variables for a gubernatorial election year and years before and after it, which were expected to capture the incumbent governor's reelection motive. Furthermore, we included explanatory variables reflecting the linkage with the federal government, because good relations between a local government and the federal government might help the local government gain large grants from the federal government, leading to larger expenditure on a program. To account for these effects, a dummy variable was constructed that takes 1 if a governor belongs to the same party as the President; it takes a value of 0 otherwise. We also used the percentage of state legislators who belong to the same party as the majority of members in the US House of Representatives, and the percentage of state legislators of the same party as the President. The estimated coefficients of these variables, however, were not statistically significant in regressions with spending on education and spending on highways as the dependent variables though the estimates are not reported here. These results indicate that the electoral motive and the influence from the federal government are weak in the state governments' choice of budgets allocated on investment in human and physical capital.

Other control variables include: unemployment rate in the state (the data are from the American Community Survey, US Census Bureau); real GDP per capita by state (in thousands of dollars, from the Regional Economic Accounts by Bureau of Economic Analysis, US Department of Commerce); real personal income per capita (in thousands of dollars), nominal personal income per capita (from the American Community Survey, US Census Bureau) deflated by the Historical Consumer Price Index for All Urban Consumers (CPI-U), US city average, Bureau of Labor Statistics (CPI-U in 2005 set as 100). We also include the dummy variable for 2008 to capture the effect of the subprime mortgage crisis.

The spending policies we consider are "education," "highways," and "public welfare." According to the definition by the US Census Bureau,[6] "education" refers to schools, colleges, and other educational institutions. Although other educational institutions and educational programs for adults, veterans, and other special classes are also included, we expect that the ratio of total spending on education would be smaller the greater the fraction of elderly in a state. "Highways" includes construction, maintenance, and operation of highways, streets, and related structures, such as bridges, tunnels, ferries, and street lighting. The expenditure on highways will generate long-term benefits to residents who will be still alive in the district. Lastly, "Public welfare" refers to support of and assistance to needy persons contingent upon their need. It includes cash assistance under the categorical programs Old Age Assistance and Temporary Assistance for Needy Families (TANF), and under any other welfare programs, and provision and operation by the government of welfare institutions, but i t excludes pensions to former employees and other benefits not contingent on need. Nursing homes are included in "public welfare" unless they are directly associated with a government hospital. We expect that the share of spending on welfare programs would be higher in states with a larger elderly population.

[6] The definitions are on https://www.census.gov/govs/definitions/index.html#p.

The elderly population may be endogenous, because much spending that benefits the elderly may attract older residents from other states, while taxes used to benefit the young may induce out-migration. For example, a state spending little on programs that mostly benefit young residents, such as education, will attract elderly residents and discourage families with young children from living in this state.

The logic provided by Tiebout (1956) indicates that with various tax-expenditure bundles provided by regional governments, individuals can choose places to live in that are best suited to their preferences. Thus, people's voting by their feet may cause migration but not a change in the tax-expenditure bundle provided by the government. Depending on this view, our regression results indicate how Tiebout (1956) sorting would mitigate a decline in public investment in human and physical capital.

This spending-inducing migration suggests that the size of the elderly population is itself a function of the spending structure in a state. Because the size of the elderly population is correlated with the unobservable determinants of the dependent variable, estimation results are likely to be biased.

To address this endogeneity, we use 10-year lagged annual average temperature of the state as an instrument for the percent elderly residents. In the US, the elderly tend to move from states with cold weather and much snow to warm states. Figure 5.1 depicts the relationship of a percent aged 65 and over in 2017 and a 10-year lagged annual average temperature by state. Alaska is the coldest state in the US; its ratio of elderly residents is the second lowest at 11%. Florida, which has the highest ratio of

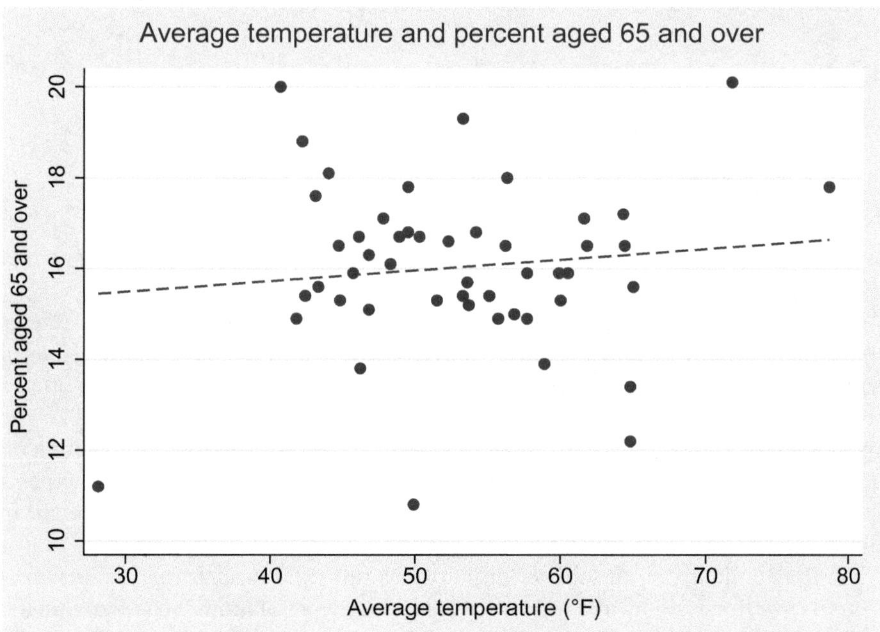

Fig. 5.1 Average temperature and percent aged 65 and over

elderly residents (20%) among all states, has the second highest average temperature (the top-ranked state is Hawaii). Indeed, the dotted line displays a positively sloped relation: the elderly are attracted to warm states. The data on annual average temperature by state come from the National Centers for Environmental Information. The statewide data for Hawaii are unavailable, so that we use city-level annual average temperature of Honolulu International Airport from 1995 to 2007.

Summary statistics for all the variables are presented in Table 5.1. The average share of citizens aged 65 and over, across states through the sample periods, is 13.9%, which is smaller than 21.9% in Japan (see Table 2.1 in Chap. 2), although the sample periods of both countries are not identical. The coefficient of variation for the US states is 0.15, which is smaller than 0.19 in Japan. Table 5.1 also shows interregional heterogeneity in racial composition. The smallest percent of Black is 0.3% (Montana in 2009); the largest is 38.0% (Mississippi in 2017). Also, percent Hispanic or Latino varies from 0.6% (West Virginia in 2005) to 48.8% (New Mexico in 2017). These differences are larger than the differences for percent aged 65 and over, which ranges from 6.6% (Alaska in 2005) to 20.1% (Florida in 2017).

Table 5.1 also shows how state governments allocate their resources across programs. On average, the highest share of budgets is spent on education. The average expenditure ratio on public welfare is ranked as the second. The average percent investment in highways is smaller than the average percent investment in human capital, although the coefficient of variation for the former (0.36) is greater than the coefficient of variation for the latter (0.17), indicating that the share of investment in highways more greatly varies across states.

5.3.2 Results

Spending on education

First, we examine how the elderly share of the population and political factors affect a government's budget allocation on education. Table 5.2 shows the results.

We conducted regression based on the fixed-effects model and the random-effects model, to consider an individual effect of each state. The value of the Sargan-Hansen statistic to test fixed-effects versus random-effects models supports the fixed-effects model, so that the results of it are presented in the second column of Table 5.2. The higher the share of the elderly in the population the smaller the share of spending on education; this effect is statistically significant. This result supports our hypothesis discussed in Chap. 1; a larger elderly population induces less public investment in human capital.

The third column presents the results of the instrumental variable regression which uses a fixed-effects model and instruments the elderly population. We still obtain the result indicating that the elderly population does not support higher spending share on education. The estimated coefficient indicates that if the share of the elderly changes by one standard deviation, the spending share on education changes in the

Table 5.1 Summary statistics

Variable	Mean	Std. dev	Minimum	Maximum
Dependent variables				
Percent expenditure on education	31.0398	5.2497	18.2857	43.7674
Percent expenditure on highways	6.8166	2.4656	2.3101	18.0650
Percent expenditure on public welfare	24.0922	4.7641	10.0460	38.7763
Explanatory variables				
Percent aged 65 and over	13.9011	2.0921	6.6	20.1
Percent female elderly	56.2948	1.7263	49.9	59.8
Percent under 15	19.5014	1.8017	15.4	26.8
Percent Black	10.4355	9.5491	0.3020	38.0126
Percent Hispanic or Latino	10.6311	10.1015	0.5723	48.7696
Percent unemployed	7.1303	2.2403	2.6	15.1
Real GDP per capita	50.1481	9.7474	32.770	79.894
Real personal income per capita	23.8923	3.5126	16.898	33.949
Democratic governor	0.4520	0.4980	0	1
Percent state Democratic legislators	48.6513	16.7290	13.3333	92.1053
Instrument				
Average temperature 10-year lag	52.6951	9.1136	24	79.1
Observations	637			

Note Data are based on 49 US states between 2005 and 2017

Data sources

Percent expenditure on education, percent expenditure on highways, percent expenditure on public welfare

The values of these variables are calculated using the data on total expenditure by function of the Annual Survey of State and Local Government Finances, US Census Bureau

Percent aged 65 and over, percent female elderly, percent under 15, percent Black, percent Hispanic or Latino

We use estimates of the American Community Survey, US Census Bureau, for selected categories

Percent unemployed

We use estimates of unemployment rate by the American Community Survey for population 16 years and over

Real personal income per capita

We use estimates of per capita income by the American Community Survey, deflated by the Historical Consumer Price Index for All Urban Consumers (CPI-U), US city average, Bureau of Labor Statistics, setting CPI-U in 2005 to 100

Real GDP per capita

The data come from the Regional Economic Accounts by Bureau of Economic Analysis, US Department of Commerce

Democratic governor

We use the data collected by the National Governors Association. We put 1 for Democratic governors, 0 otherwise. When a governor is replaced by another governor in the mid of year, a weighted sum of values, on the basis of the length of incumbency, is put in as a value

Percent state Democratic legislators

The data are constructed using the data collected by the National Conference of State Legislatures

Table 5.1 (continued)

Annual average temperature
The data come from the National Centers for Environmental Information. The statewide data for Hawaii are not available, so that we used city-level average temperature data for Honolulu International Airport in Hawaii, for 1995 to 2007
We exclude Nebraska from our analysis because Nebraska's legislature is unicameral and nonpartisan, differently from other states
Definition
We refer to https://www.census.gov/govs/definitions/index.html#p
• **Education**—Schools, colleges, and other educational institutions (e.g., for blind, deaf, and other handicapped individuals), and educational programs for adults, veterans, and other special classes. State institutions of higher education includes activities of institutions operated by the state, except that agricultural extension services and experiment stations are classified under Natural resources and hospitals serving the public are classified under Hospitals. Revenue and expenditure for dormitories, cafeterias, athletic events, bookstores, and other auxiliary enterprises financed mainly through charges for services are reported on a gross basis
• **Highways**—Construction, maintenance, and operation of highways, streets, and related structures, including toll highways, bridges, tunnels, ferries, street lighting and snow and ice removal. However, highway policing and traffic control are classed under Police protection
• **Public welfare**—Support of and assistance to needy persons contingent upon their need. Excludes pensions to former employees and other benefits not contingent on need. Expenditures under this heading include: Cash assistance paid directly to needy persons under the categorical programs Old Age Assistance, Temporary Assistance for Needy Families (TANF) and under any other welfare programs; Vendor payments made directly to private purveyors for medical care, burials, and other commodities and services provided under welfare programs; and provision and operation by the government of welfare institutions. Other public welfare includes payments to other governments for welfare purposes, amounts for administration, support of private welfare agencies, and other public welfare services. Health and hospital services provided directly by the government through its own hospitals and health agencies, and any payments to other governments for such purposes are classed under those functional headings rather than here
• **Hospitals**—Financing, construction acquisition, maintenance or operation of hospital facilities, provision of hospital care, and support of public or private hospitals. Own hospitals are facilities administered directly by the government concerned; Other hospitals refers to support for hospital services in private hospitals or other governments. However, see Public welfare concerning vendor payments under welfare programs. Nursing homes are included under Public welfare unless they are directly associated with a government hospital
• **Health**—Outpatient health services, other than hospital care, including: public health administration; research and education; categorical health programs; treatment and immunization clinics; nursing; environmental health activities such as air and water pollution control; ambulance service if provided separately from fire protection services, and other general public health activities such as mosquito abatement. School health services provided by health agencies (rather than school agencies) are included here. Sewage treatment operations are classified under Sewerage

opposite direction by 4.1%. For instance, if the population share of the elderly in Florida declined from 20.1% to the state average 13.9%, then the relative spending on education would increase by 12.2%. If the state average of the ratio of the elderly in the US, through sample periods, increased from 13.9% up to the 2015 level of Japan, 26%, the average relative spending on education would decline by 23.7%.

The fourth column presents the estimated coefficient on the instrument in the first-stage regression. It reports that the annual average temperature 10 years before in a

Table 5.2 Effects of the elderly population on expenditure on education

Dependent variable	Percent expenditure on education(FE)	Percent expenditure on education(FE-IV)	Percent aged 65 and over (First-stage)
Average temperature 10-year lag			0.08395*** (0.02796)
Percent aged 65 and over	−0.69436** (0.27554)	−1.95990*** (0.67337)	
Percent under 15	−0.04061 (0.30576)	−1.15169 (0.71857)	−0.86845*** (0.13409)
Percent Black	0.39776 (0.40974)	0.53180 (0.52094)	0.10671 (0.16388)
Percent Hispanic or Latino	0.22432 (0.25575)	0.84867** (0.33251)	0.46931*** (0.09323)
Percent unemployed	−0.35568*** (0.09315)	−0.46077*** (0.09151)	−0.10052*** (0.02951)
Real GDP per capita	−0.20069*** (0.03881)	−0.25696*** (0.04567)	−0.04674*** (0.01673)
Real personal income per capita	0.16751 (0.20323)	0.41792 (0.27775)	0.17212*** (0.05934)
Democratic governor	0.45932* (0.27295)	0.37149 (0.31867)	−0.05410 (0.14180)
Percent state Democratic legislators	0.02083 (0.01898)	−0.00564 (0.02254)	−0.02023*** (0.00734)
2008 dummy	0.26627 (0.17416)	−0.17865 (0.30228)	−0.51454*** (0.07961)
R^2 within between overall	0.2796 0.0357 0.0438	0.1346 0.0024 0.0044	0.8388 0.0000 0.0170
Sargan-Hansen statistic	Chi-sq(10) = 28.119 P-value = 0.0017		
F value			F(10,48) = 103.79 Prob > F = 0.0000
Observations	637	637	637

Note FE means fixed-effects model and IV means instrumental variable regression. Cluster-robust standard errors are in parentheses. Estimation uses data from 49 states between 2005 and 2017. ***Significant at 1% level, **Significant at 5% level, *Significant at 10% level

state is associated with an increase in the share of the elderly citizens. This relation is consistent with the graph in Fig. 5.1. Thus, our proposed instrument is valid.

We shall further examine the results of the instrumental variable method. It is remarkable that the population ratio under 15 years old is not associated with a higher spending ratio on education. Rather, race has a stronger effect. An increase in percent Hispanic or Latino population by one standard deviation would increase the share of total spending on education by 8.6%.

A decrease in the unemployment rate and an increase in real personal income per capita increase spending on education, though the latter effect is not significant. This relation is consistent with the previous literature that uses aggregate data and examines income effect on education (Ladd 1975; Lovell 1978), indicating that education is a normal good. However, an increase in real GDP per capita reduces the share of public spending on education; good business conditions may induce the government to spend more on other programs. Democratic governors' positive effect on relative spending on education, which was significant without instrumenting the ratio of the elderly citizens, is insignificant in the instrumental variable analysis.

Spending on highways

Table 5.3 presents estimates of relative spending on highways as the dependent variable. The middle column is associated with the panel data analysis using a fixed-effects model, and the rightmost column is for a regression that uses a fixed-effects model and instruments the elderly population.[7] In the analysis that addresses the endogeneity of the population share of the elderly, we obtain the estimated coefficient of the population ratio of the elderly which is not statistically significant, although it is still negative. This contrasts with the result for relative spending on education in Table 5.2, and even with the result on public investment in Japan (see Chap. 2). This may be caused by citizens' mobility in the US. Young individuals, who may migrate in the future, would not like to invest in local infrastructure, but would rather invest in human capital. Thus, the effect of the elderly population on relative spending on highways is weak. We cannot find that a larger elderly population reduces public investment in physical capital.

In Table 5.3 effects of racial composition appear to be weak. Among economic variables, higher real GDP per capita increases the expenditure ratio on highways. The subprime loan crisis reduced it. We cannot find significant influences of political factors.

Comparison with the relative spending on public welfare

We analyze relative spending on public welfare, to compare the effect of the elderly population on public investment with its effect on redistributive policies.[8] Table 5.4 reports the results of instrumented fixed-effects model analysis for relative spending on public welfare. The estimated coefficient, without using the instrument for the elderly population, is positive and statistically significant, but by employing instrumental variable method, we obtain a different result. A higher share of the elderly population has a negative but statistically insignificant effect. This finding contradicts our prediction but is consistent with Parsons (1982) who finds that the percentage of the population over 65 had no effect on state Old Age Assistance benefits during the 1930s. He interprets this as a negative effect of the increased burden on younger voters of making transfers to numerous elderly recipients, offsetting a positive effect

[7] The value of the Sargan-Hansen statistic in Table 5.3 favors the fixed-effects model.

[8] The test on fixed-effects versus random-effects models supports the adoption of fixed-effects model, as reported in Table 5.4.

Table 5.3 Effects of the elderly population on expenditure on highways

Dependent variable	Percent expenditure on highways(FE)	Percent expenditure on highways(FE-IV)
Percent aged 65 and over	−0.28989* (0.15460)	−0.59189 (0.45859)
Percent under 15	0.06576 (0.15805)	−0.19939 (0.40922)
Percent Black	−0.01117 (0.17755)	0.02082 (0.20516)
Percent Hispanic or Latino	−0.00110 (0.15007)	0.14789 (0.24442)
Percent unemployed	−0.03253 (0.04947)	−0.05761 (0.05182)
Real GDP per capita	0.09265* (0.04772)	0.07922* (0.04466)
Real personal income per capita	0.03429 (0.08243)	0.09405 (0.11889)
Democratic governor	0.02243 (0.16743)	0.00147 (0.18076)
Percent state Democratic legislators	0.00021 (0.01137)	−0.00610 (0.01366)
2008 dummy	−0.25518* (0.13738)	−0.36136* (0.21752)
R^2 within between overall	0.1790 0.0018 0.0083	0.1560 0.0546 0.0302
Sargan-Hansen statistic	Chi-sq(10) = 24.816 P-value = 0.0057	
Observations	637	637

Note Estimates are based on fixed-effects model. IV means instrumental variable regression. Cluster-robust standard errors are in parentheses. Estimation uses data from 49 states between 2005 and 2017. ***Significant at 1% level, **Significant at 5% level, *Significant at 10% level

of the increased voting power of the elderly. And it is noticeable that a larger share of Black and a larger share of Hispanic or Latino in the state do not have strong effects on the spending ratio on public welfare.

It is puzzling that a lower unemployment rate is associated with an increase in the share of spending on public welfare, because public welfare includes support contingent upon need. It is also remarkable that having the Democratic governor or having a higher ratio of Democratic legislators is not explicitly linked with an increase in the share of spending on public welfare.

Table 5.4 Effects of the elderly population on expenditure on public welfare

Dependent variable	Percent expenditure on public welfare(FE)	Percent expenditure on public welfare(FE-IV)
Percent aged 65 and over	0.96125*** (0.31446)	−0.40929 (2.17340)
Percent under 15	−0.15205 (0.42139)	−1.35532 (1.92593)
Percent Black	−0.14948 (0.50796)	−0.00431 (0.56868)
Percent Hispanic or Latino	0.12198 (0.27183)	0.79813 (1.12214)
Percent unemployed	−0.34497*** (0.08506)	−0.45878** (0.21607)
Real GDP per capita	−0.12087 (0.08973)	−0.18180 (0.14849)
Real personal income per capita	0.05763 (0.22065)	0.32882 (0.51121)
Democratic governor	0.16395 (0.38444)	0.06884 (0.44501)
Percent state Democratic legislators	−0.02837 (0.02366)	−0.05703 (0.05587)
2008 dummy	−1.21322*** (0.19840)	−1.69504** (0.86378)
R^2 within between overall	0.5097 0.0397 0.1077	0.4266 0.0057 0.0184
Sargan-Hansen statistic	Chi-sq(10) = 91.114 P-value = 0.0000	
Observations	637	637

Note Estimates are based on fixed-effects model. IV means instrumental variable regression. Cluster-robust standard errors are in parentheses. Estimation uses data from 49 states between 2005 and 2017. ***Significant at 1% level, **Significant at 5% level, *Significant at 10% level

Robustness check

We then use governmental spending per capita on highways and public welfare as a dependent variable, to examine whether estimation results are robust. Influences of elderly citizens are qualitatively similar to the results obtained from the estimation employing percent expenditure as a dependent variable, though we do not present here estimation results on the per-capita basis. In fixed-effects model regression without instrumenting percent elderly, expenditure per capita on highways declines with the elderly ratio in a state, while a higher share of the elderly population increases expenditure per capita on public welfare; the effects of the elderly ratio are insignificant in instrumental variable analysis. For public expenditure on education, we will construct another data set on the per-pupil basis and continue to examine in Sect. 5.4.

5.4 Per Pupil Public Expenditure on Education

As reported in Table 5.2, the fraction of governmental spending on education declines with the fraction of the population that is elderly. How are the results modified if we adopt a different measure for public spending on education? Poterba (1997, 1998) used per child school spending that includes outlays by state and local governments, as a dependent variable, and found that a growing elderly population lowered education spending in his panel data analysis allowing for state and time fixed effects. He uses the data on education spending in each continental state of the US from fiscal years 1961, 1971, 1981, and 1991. Miller (1996), using the decennial data from 1960 to 1990, confirms a positive but statistically insignificant effect of the elderly share on per adult state and local spending for public education. The more recent work by Harris et al. (2001) studies the impact of an aging population on revenues per pupil that are spent on public education in the US. In the analysis with the data of public-school districts, the elderly population has only a modest negative effect on education spending. However, they confirm that a larger share of the elderly at the state level tends to depress state spending on education. The authors consider these results as generated from the elderly's belief that only higher local spending is capitalized into house values.

We estimate regressions with per pupil current spending and per pupil capital outlay for public elementary and secondary education as dependent variables. The reason we employ these two categories of expenditure each in a regression model is that capital outlay generates more long-lived benefits than current spending.[9] Current spending includes salaries and wages, and employee benefits, for both instruction and support services. Capital outlay includes direct expenditure for construction of buildings, roads, and other improvements, for purchases of equipment, land, and existing structures, and for payments on capital leases. This includes amounts for additions, replacements, and major alterations to fixed works and structures, but expenditure for maintenance and repairs to such works and structures is classified as current spending.

In estimation, we have to consider effects of the elderly share which may work in opposite directions. First, increased voting power of the elderly citizens may induce governments to reduce per pupil expenditure for education, in particular, per pupil capital outlay. On the other hand, the elderly might support public education because they may be altruistic, or because good education in a region may attract young workers, so lowering prices for services the elderly consume, as stated in Sect.1.3. Furthermore, higher spending on education may be capitalized into the value of houses and land, which will benefit elderly homeowners (Harris et al. 2001).

Another factor that should be considered is people's voting by their feet. If the elderly vote with their feet in the manner Tiebout (1956) suggests, they would move from a region with high educational spending to a region with low educational spending. Then, even if an elderly population does not support expenditure

[9] Poterba (1997, 1998) and Harris et al. (2001) use total government revenues or expenditure in the calculation of the dependent variable.

on public education, Tiebout (1956) sorting by the elderly could leave education spending unchanged, simply inducing migration (Harris et al. 2001). In regard to the selection of the estimation method, we instrument percent aged 65 and over with the annual average temperature 10-year lag, as in Sect. 5.3, to address the endogeneity of that variable.

The data of annual spending on public elementary and secondary school system (in dollars) are from Public Elementary-Secondary Education Finance Data, US Census Bureau. This reports public school system finances on a fiscal-year basis that starts in July in the previous year and ends in June in many states. Therefore, we adopt one-fiscal-year lead data for the dependent variable, to match explanatory variables that are on a calendar-year basis, deflating by the Historical Consumer Price Index for All Urban Consumers (CPI-U), US city average, Bureau of Labor Statistics (CPI-U in 2005 set as 100). The data of per pupil current spending exclude expenditure for adult education, community services, and other nonelementary-secondary programs. The data on capital outlay are given only in the total amount, so that we calculate per pupil capital outlay by dividing capital outlay in a state by the number of the elementary-secondary enrollment. Figure 5.2 displays the relation between the elderly ratio and per pupil current spending/capital outlay for 48 continental states in fiscal year 2017. It appears to support our hypothesis that the elderly do not support public expenditure used as capital outlay.

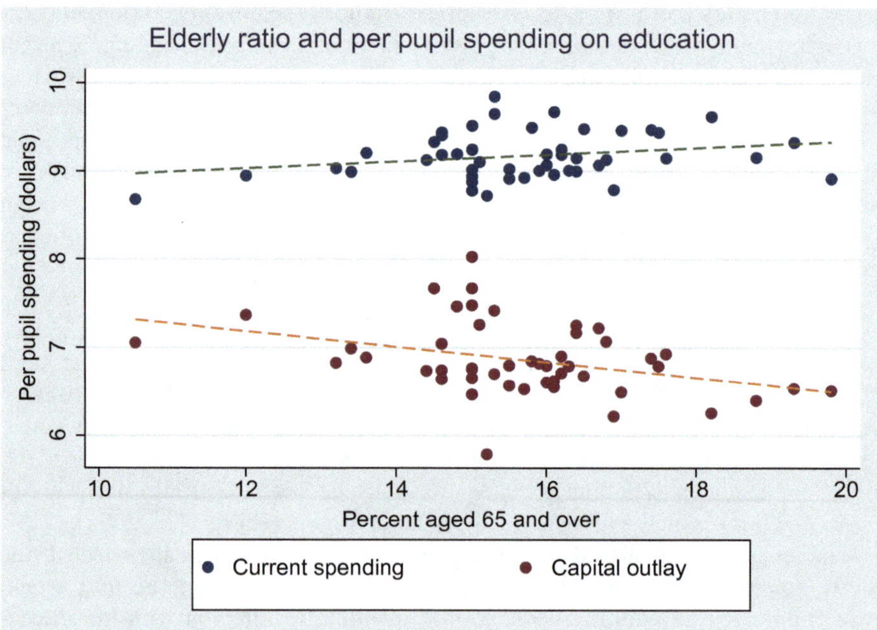

Fig. 5.2 Elderly ratio and per pupil spending on education

Table 5.5 Summary statistics of selected variables

Variable	Mean	Std. dev	Minimum	Maximum
Dependent variables				
Per pupil current spending	9588.586	2463.339	5418.996	18,935.7
Per pupil capital outlay	1024.089	507.740	133.395	4014.41
Explanatory variables				
Percent aged 5–14	13.1083	1.063	10.6	17.5
Homeownership rate	68.8616	4.9698	51.5	81.325
Percent nonwhite	20.6995	10.2309	3.3897	43.5339
Per pupil federal resources	1033.703	297.017	567.1954	2686.688
Observations	576			

Note Data are based on 48 US states between 2005 and 2016

In choice of explanatory variables, we closely follow the regression model by Poterba (1997). Besides population share aged 65 and over and real personal income per capita used in Sect. 5.3, real federal revenue per pupil is included. The data on it are obtained from the same source as spending on education. It is expected to raise per pupil spending. We also include homeownership rate as an explanatory variable. Homeowners in a region likely favor higher spending on education than the residents who live in rented houses, because higher quality of education in a region will be capitalized into the value of houses and land they own. The data are from Current Population Survey/Housing Vacancy Survey, US Census Bureau. Moreover, we include the percentage of population from 5 to 14 years old, which almost covers the elementary and secondary school ages, and the percentage of population nonwhite, the data of which come from American Community Survey. The summary statistics of added variables are displayed in Table 5.5.

Observations are from 48 continental states in the US between 2005 and 2016. Following Poterba (1997, 1998), all of the variables are measured in logarithms, so the estimated coefficients can be interpreted as elasticities.

The results by uninstrumented fixed-effects model regression are in Table 5.6. The fourth column in it is comparable to the results of Poterba (1997). The estimated elasticity of per pupil total spending on education with respect to the elderly population share (−0.076) is smaller than the estimated elasticity in Poterba (1997) (−0.276), and statistically insignificant. This may be caused from using annual data of shorter sampling periods, differently from Poterba (1997). Remarkably, a higher share of the elderly citizens has a stronger effect on capital outlay than on current spending, which supports our hypothesis that the elderly would not support public spending that generates long-lasting benefits. Also, income effects on the demand for educational spending and the effect of federal resources are apparent.

Table 5.7 is for instrumental variable regression, which Poterba (1997, 1998) did not conduct. Differently from the results in Table 5.6, with the effects of personal income and federal resources controlled, the decline of per pupil capital outlay caused

Table 5.6 Effects of the elderly population on per pupil spending on education

Dependent variable	Log (real current spending per pupil) (FE)	Log (real capital outlay per pupil) (FE)	Log (real total spending per pupil) (FE)
Log (percent aged 65 and over)	−0.01056 (0.07852)	−0.84590* (0.47418)	−0.07599 (0.10390)
Log (percent 5 to 14 years old)	−0.51782*** (0.19159)	0.51532 (0.88975)	−0.44775** (0.21170)
Log (homeownership rate)	0.17426 (0.11869)	0.65692 (1.0605)	0.30254 (0.19764)
Log (percent nonwhite)	0.20245*** (0.05144)	0.30843 (0.26592)	0.22125*** (0.06729)
Log (real personal income per capita)	0.63294*** (0.06113)	4.24680*** (0.45506)	1.00627*** (0.10008)
Log (real federal resources per pupil)	0.08218*** (0.00918)	0.19517** (0.07571)	0.09240*** (0.01335)
R^2 within between overall	0.4179 0.2538 0.2592	0.4249 0.0655 0.0935	0.4627 0.3148 0.3209
Sargan-Hansen statistic	Chi-sq(6) = 52.338 P-value = 0.0000	Chi-sq(6) = 69.258 P-value = 0.0000	Chi-sq(6) = 29.248 P-value = 0.0001
Observations	576	576	576

Note Estimates are based on fixed-effects model. Cluster-robust standard errors are in parentheses. Estimation uses data from 48 states between 2005 and 2016. ***Significant at 1% level, **Significant at 5% level, *Significant at 10% level

by a larger share of elderly population is no longer detected. This indicates that Tiebout (1956) sorting works well. Reasoning is as follows. Some states spend much on education per pupil, and other states spend little. If Tiebout (1956) mechanism is working, the elderly move to states with low education spending. Then, in a regression without instrumenting the elderly ratio, we should find a negative coefficient on it. But if we instrument the elderly ratio, we should find no impact of this variable, which is what the data show.

5.5 Gender Effects

As demonstrated in Sect. 5.1, the male elderly are, on average, more highly educated than the female elderly. And the life expectancy for women is longer than for men. In considering influences on politics by gender difference, these two features make the effects of the elderly population depend on the gender composition in a state. If less

Table 5.7 Effects of the elderly population on per pupil spending on education (instrumental variable regression)

Dependent variable	Log (real current spending per pupil) (FE-IV)	Log (real capital outlay per pupil) (FE-IV)	Log (real total spending per pupil) (FE-IV)
Log (percent aged 65 and over)	0.33943 (0.33723)	0.86038 (1.48556)	0.48685 (0.35884)
Log (percent 5–14 years old)	−0.24465 (0.33445)	1.84707 (1.64064)	−0.00845 (0.36952)
Log (homeownership rate)	0.65714 (0.42894)	3.01107 (2.23755)	1.07909** (0.47059)
Log (percent nonwhite)	0.10719 (0.09337)	−0.15602 (0.43232)	0.06805 (0.10562)
Log (real personal income per capita)	0.65604*** (0.09485)	4.35938*** (0.45641)	1.04340*** (0.14084)
Log (real federal resources per pupil)	0.09334*** (0.01424)	0.24954** (0.10271)	0.11034*** (0.01954)
R^2 within between overall	0.3358 0.4940 0.4812	0.3687 0.0181 0.0466	0.3275 0.5113 0.4984
Observations	576	576	576

Note Estimates are based on fixed-effects model. IV indicates instrumental variable regression. Cluster-robust standard errors are in parentheses. Estimation uses data from 48 states between 2005 and 2016. ***Significant at 1% level, **Significant at 5% level, *Significant at 10% level

educated individuals have weaker influence on politics, a larger elderly female population likely weakens a negative effect of an elderly population on public investment in education. And if they count benefits from public investment for a long while, they may not favor decreasing public investment.

To capture gender effects, we construct a new variable, percent female elderly, which is the percent female in the population aged 65 and over (the summary statistics on the percent female elderly are reported in Table 5.1). The estimates using this variable are in Table 5.8. The estimated effect of percent female elderly on percent expenditure on education is negative, though statistically insignificant. We also obtain negative effects of percent female elderly for percent expenditure on highways, which are not reported in the table. These results can be interpreted as indicating that low education does not weaken the elderly's political influence.

5.6 Conclusion

We conducted empirical analyses on the effects of the elderly on budget allocation by states in the US. We hypothesized that a larger elderly population likely induces the government to adopt policies they prefer, instead of policies that mainly benefit

Table 5.8 Gender effects

Dependent variable	Percent expenditure on education(FE-IV)
Percent aged 65 and over	−2.88416** (1.15823)
Percent female elderly	−1.19262 (0.76827)
Percent under 15	−1.65251* (0.97188)
Percent Black	0.59076 (0.55542)
Percent Hispanic or Latino	0.79317** (0.31107)
Percent unemployed	−0.62131*** (0.14853)
Real GDP per capita	−0.32777*** (0.07641)
Real personal income per capita	0.37143 (0.25913)
Democratic governor	0.33017 (0.36162)
Percent state Democratic legislators	0.01978 (0.02376)
2008 dummy	−0.07093 (0.25802)
R^2 between overall	0.0073 0.0101
Observations	637

Note Estimates are based on fixed-effects model. IV means instrumental variable regression. Cluster-robust standard errors are in parentheses. Estimation uses data from 49 states between 2005 and 2017. ***Significant at 1% level, **Significant at 5% level, *Significant at 10% level

the young. In particular, as the elderly share increases, the share of spending on education and infrastructure would be reduced.

It is noteworthy that this expected effect of an elderly population is more evident for education, in comparison with its effect on public investment in infrastructure. The benefits generated by education are expected to be long lasting in the society. They would spill over generations and across regions, so that reduced spending on education could hurt long-run growth. However, the results of our additional analysis on education spending on the per-pupil basis imply that the effects of large voting power of the elderly might be mitigated by voting by their feet. Remarkably, the reduced effects are apparent not only for current spending but also for capital outlay in elementary and secondary school finance, once the effect of migration is considered. This indicates that fiscal decentralization which allows heterogenous communities with homogenous citizens in each of them might be a solution in the aged society.

References

Alt JE, Lowry RC (1994) Divided government and budget deficits: evidence from the states. Am Political Sci Rev 88(4):811–828
Besley T, Case A (2003) Political institutions and policy choices: evidence from the United States. J Econ Lit 41(1):7–73
Garand JC (1988) Explaining government growth in the U.S. states. Am Political Sci Rev 82(3):837–849
Gilens M (1996) "Race coding" and white opposition to welfare. Am Political Sci Rev 90(3):593–604
Harris AR, Evans WN, Schwab RM (2001) Education spending in an aging America. J Public Econ 81(3):449–472
Hawes DP, McCrea AM (2018) Give us your tired, your poor and we might buy them dinner: social capital, immigration, and welfare generosity in the American States. Political Res Q 71(2):347–360
Hettich W, Winer SL (1988) Economic and political foundations of tax structure. Am Econ Rev 78(4):701–712
Hettich W, Winer SL (1999) Democratic choice and taxation: a theoretical and empirical analysis. Cambridge University Press, New York
Ladd H (1975) Local education expenditures, fiscal capacity, and the composition of the property tax base. National Tax J 28(2):145–158
Lovell M (1978) Spending for education: the exercise of public choice. Rev Econ Stat 60(4):487–495
Metzler A, Richard S (1981) A rational theory of the size of government. J Political Econ 89(5):914–927
Miller C (1996) Demographics and spending for public education: a test of interest group influence. Econ Educ Rev 15(2):175–185
Parsons D (1982) Demographic effects on public charity to the aged. J Hum Res 17(1):144–152
Poterba JM (1997) Demographic structure and the political economy of public education. J Policy Anal Manage 16(1):48–66
Poterba JM (1998) Demographic change, intergenerational linkages, and public education. Am Econ Rev 88(2):315–320
Roberts KWS (1977) Voting over income tax schedules. J Public Econ 8(3):329–340
Rogers DL, Rogers JH (2000) Political competition and state government size: do tighter elections produce looser budgets? Public Choice 105(1/2):1–21
Romer T (1975) Individual welfare, majority voting, and the properties of a linear income tax. J Public Econ 4(2):163–185
Soss J, Fording RC, Schram SF (2008) The color of devolution: race, federalism, and the politics of social control. Am J Political Sci 52(3):536–553
Tiebout C (1956) A pure theory of local expenditures. J Political Econ 64(5):416–424

Chapter 6
The Effect of the Elderly on Taxation and Minimum Wages in the US States

6.1 Introduction

Consider individual labor supply and consumption from the life-cycle perspective. Because many of the elderly are retired, they may little directly benefit from state policies that increase wage income. But they may benefit indirectly: lower taxes on labor may induce in-migration of the young (or less out-migration by the young), which may consequently increase labor supply and reduce the price of services the elderly buy, for instance, nursing care. Because a higher tax burden on local businesses may lead to price increases, the elderly may prefer a small corporate income tax. And for similar reasons they may favor a low minimum wage.

In this chapter we will test whether the elderly care about the burden imposed by the state governments on labor and businesses. In particular, we will estimate regression models that incorporate strategic interdependence among state governments. Each state government may react to other governments' choices. The US provides a laboratory for investigating governmental competition because, while states have authority to choose their own policies, they share many important institutional and political features.

Governments have to consider two factors in competing against each other. The first incentive is to attract capital, which may be accomplished by having a small corporate income tax and a small minimum wage. The second concern is on labor. In Sect. 1.3 we argued that increased public spending on goods that working-aged people value can increase labor supply, by reducing out-migration of such people, or inducing in-migration. But an increase in the minimum wage would generate a different result. A higher minimum wage should be transferred to the price of goods and services, thereby increasing the price of goods and services produced by low-wage labor.[1]

[1] Using the US data, Cadena (2014) and Monras (2019) show that low-skilled labors tend to leave or avoid moving to the regions that increase minimum wage.

K. Terai et al., *The Political Economy of Population Aging*, Advances in Japanese Business and Economics 30, https://doi.org/10.1007/978-981-16-5536-4_6

Fig. 6.1 Marginal factor cost facing a monopsonist

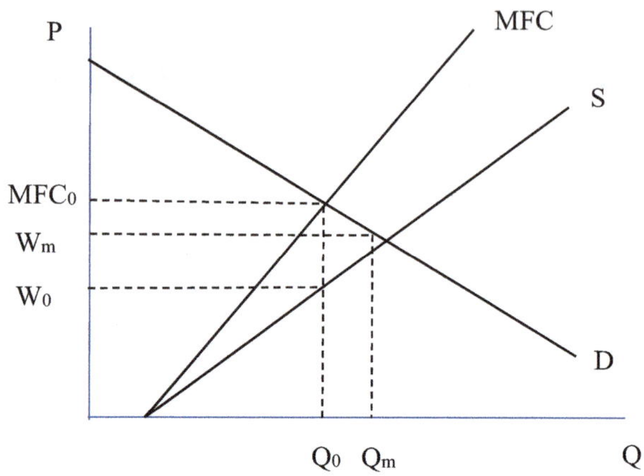

One exception may occur if employers have monopsony power. A minimum wage in that case can reduce the marginal cost of labor (because increased employment does not increase the wage), and so increase the supply of goods and services produced by such labor. The effect is shown in Fig. 6.1. An upward sloping supply curve of labor facing a monopsonist is marked by *S*. The more workers the firm hires, the higher wage it must pay to all workers. So the added cost of hiring a worker, which is called the marginal factor cost (MFC), exceeds the wage it pays: the MFC curve lies above the supply curve. Given the firm's demand for workers, which is marked by *D*, it will hire Q_0 workers, at a wage W_0.

Now consider a minimum wage, W_m. The firm pays W_m for each additional worker, and hiring additional workers does not increase the wage paid for its other workers. So the firm will hire Q_m workers. A higher employment with the minimum wage can increase the supply of goods and services, lowering prices. Thus, if monopsony prevails in the labor market for low-wage workers, the elderly may not oppose the government's intervention into the labor market by use of a minimum wage.

6.2 Related Literature

Strategic interactions of fiscal policies among governments have been well discussed in both theoretical and empirical studies.[2] In comparison with tax competition and environmental policy competition, the empirical literature on regulatory competition is small. Some examples include Davies and Vadlamannati (2013). Their hypothesis that under globalization nations compete for investment by relaxing labor standards to attract firms is confirmed by panel data for 135 countries. They find that the labor standards in one country are positively correlated with those elsewhere. Li

[2] For the empirical literature, see the survey by Breuckner (2003).

et al. (2019) investigate minimum wage competition in China. They examine city-level strategic interactions in setting and enforcing minimum wages, finding spatial interdependence among the main Chinese cities.

In estimating the effect of the elderly on minimum wage setting by a state, we use a linear interjurisdictional competition model that specifies the reaction function of a jurisdiction against other jurisdictions' choices. This regression model has often been used in empirical work on tax competition. For instance, Jacobs et al. (2010) analyzed consumption tax competition between US states. They find that the choices of sales taxes by the US states are positively related. These studies focus on strategic interdependence between jurisdictions and the need for policy coordination among them, rather than on the effects of demographics.

Monopsony in the low-wage labor market has gathered interest in theoretical and empirical analyses. Stigler (1946) and Card and Krueger (1995) show that with an employer having control over the wage rate and therefore facing upward-sloping labor supply curves, a minimum wage can increase employment. However, empirical evidence of monopsonistic competition is mixed. Hirsch and Schumacher (1995) find no positive relationship between relative nurse/non-nurse wage rates and hospital density or market size, indicating little support for the view that monopsony power plays an important role in nursing labor markets. The elasticity of nurses' labor supply to individual Department of Veterans Affairs hospitals, responding to changes in wages, is estimated by Staiger et al. (2010). Evidence showing inelastic labor supply suggests that hospitals are wage setters in the labor markets for registered nurses. Matsudaira (2014) estimates the firm-level elasticity of labor supply for nurse aids in the long-term care (nursing home) industry. The results suggest that firms have very little market power over the wages of nurses.

When wages are below marginal productivity, as with monopsony, employers are able to increase wages without laying off workers. Azar et al. (2019) provide empirical support for the monopsony explanation by studying a low-wage retail sector in the US. They show that more concentrated labor markets, where wages are more likely to be below marginal productivity, experience significantly more positive employment effects from the minimum wage. The minimum wage hikes are expected to cause prices to rise in competitive labor markets but potentially fall in monopsonistic environments. Aaronson et al. (2008), using store-level and aggregated Consumer Price Index data, provide the evidence that restaurant prices rise in response to minimum wage increases.

6.3 The Elderly and the Structure of Tax Revenues

First, we will examine how the structure of tax revenues in a state is affected by a higher share of elderly citizens in the total population. We estimate the following equation, like (5.1) in Chap. 5:

$$y_{it} = \alpha_i + x_{it}^{'}\beta + z_{it}^{'}\gamma + \varepsilon_{it}. \tag{6.1}$$

In (6.1), the dependent variable y_{it} is the percentage of each source of tax revenue in state i in year t; α_i denotes an individual effect for state i; x_{it} and z_{it} are vectors of demographic variables and other control variables; β and γ are vectors of parameters to be estimated; ε_{it} is an error term for state i in year t. A n individual effect should be considered because state-specific factors, such as geographic feature and industrial agglomeration, can affect local tax composition.

Dependent variables are the percentage of total tax revenue raised by the personal income tax, corporate income tax, and sales tax. Our panel data set covers 49 states from 2005 to 2016. Nebraska is excluded from our samples because the legislature of Nebraska is unicameral, differently from other states. The data source is the Annual Survey of State and Local Government Finances, US Census Bureau. Table 6.1 reports the summary statistics on these variables. We find wide variation in the composition of tax revenues across states.

Demographic variables include percent population aged 65 and over, and the total population (in one hundred thousands). The data for these demographic variables are from the American Community Survey, US Census Bureau. We focus on the estimated coefficient of the percentage of the population aged 65 and over. We expect no effect of a higher share of the elderly on the personal income tax. If, however, the elderly have sufficient foresight, they may prefer a low tax burden on labor income, because a high tax burden might induce out-migration of the young, and hence, a decline in labor supply, in the long run. There is little reason to believe that the elderly favor higher corporate income taxes and sales taxes. Rather, they may prefer lowering them, fearing higher taxes on these tax bases may inflate consumer prices.

As other control variables, we include political and economic variables that were used in the regressions in Chap. 5. We use the ratio of Democratic legislators in the state legislature, and a variable that takes 1 for Democratic governors and 0 otherwise.[3] The data on governors are obtained from the National Governors Association, and the data on state legislators are collected by the National Conference of State Legislatures. Variables capturing local economic conditions include unemployment rate (the data are from the American Community Survey, US Census Bureau); real GDP per capita (in thousands of dollars, from the Regional Economic Accounts by Bureau of Economic Analysis, US Department of Commerce); real personal income per capita (in thousands of dollars, from the American Community Survey, US Census Bureau). To obtain GDP per capita and personal income per capita in the real term, the nominal value is deflated by the Historical Consumer Price Index for All Urban Consumers (CPI-U), US city average, Bureau of Labor Statistics (CPI-U in 2005 set as 100).

The elderly share may be endogenous. We address endogeneity by using as an instrumental variable the 10-year lagged annual average temperature of the state, as in Chap. 5, because in the US, the elderly tend to move from states with bad weather

[3] If a governor is replaced by another governor in midyear, a weighted sum of these values is calculated on the basis of the length of incumbency.

Table 6.1 Summary statistics on tax revenues

Variable	Mean	Std. dev	Minimum	Maximum
Dependent variables				
Percent personal income tax revenue	30.9372	17.3977	0	72.5559
Percent corporate income tax revenue	5.6171	4.2249	−0.0004	33.0726
Percent sales tax revenue	47.3773	16.4733	3.1725	88.9474
Explanatory variables				
Percent aged 65 and over	13.7236	2.0120	6.6	19.8
Population	62.6794	68.8750	4.9523	392.5002
Percent unemployed	7.3015	2.2289	2.6	15.1
Real GDP per capita	49.9977	9.7322	32.77	79.894
Real personal income per capita	23.7843	3.4801	16.8978	33.949
Democratic governor	0.4624	0.4989	0	1
Percent state Democratic legislators	49.1418	16.5158	13.3333	90.2655
Instrument				
Average temperature 10-year lag	52.6736	9.1207	24	79.1
Observations	588			

Note Data are based on 49 US states between 2005 and 2016. We exclude Nebraska from our analysis because the legislature of Nebraska is unicameral and nonpartisan, differently from other states
Data sources
Percent personal income tax revenue, percent corporate income tax revenue, percent sales tax revenue
The data come from the Annual Survey of State and Local Government Finances, US Census Bureau
Percent aged 65 and over, population
We use estimates of the American Community Survey, US Census Bureau, for selected categories
Percentage unemployed
We use estimates of unemployment rate by the American Community Survey for population 16 years and over
Real personal income per capita
We use estimates of per capita income by the American Community Survey, deflated by the Historical Consumer Price Index for All Urban Consumers (CPI-U), US city average, Bureau of Labor Statistics, setting CPI-U in 2005 as 100
Real GDP per capita
The data come from the Regional Economic Accounts by Bureau of Economic Analysis, US Department of Commerce
Democratic governor
We use the data collected by the National Governors Association. We put 1 for Democratic governors, 0 otherwise. When a governor is replaced by another governor in the mid of year, a weighted sum of values, on the basis of the length of incumbency, is put in as a value
Percent state Democratic legislators
The data are constructed using the data collected by the National Conference of State Legislatures
Annual average temperature
The data come from the National Centers for Environmental Information. The statewide data for Hawaii are unavailable, so that we used city-level average temperature data for Honolulu International Airport in Hawaii, for 1995 to 2006

Table 6.2 Effects of the elderly population on the composition of tax revenues

Dependent variable	Percent personal income tax revenue (FE-IV)	Percent corporate income tax revenue (FE-IV)	Percent sales tax revenue (FE-IV)
Percent aged 65 and over	0.47928 (0.37068)	−0.90544** (0.44590)	−0.48640 (0.49275)
Population	0.02685 (0.06408)	0.08062 (0.06391)	0.17241** (0.07922)
Percent unemployed	−0.25422** (0.11161)	−0.46031*** (0.08704)	0.30413* (0.16425)
Real GDP per capita	−0.20952*** (0.06139)	−0.07910 (0.06774)	−0.19030 (0.16611)
Real personal income per capita	0.27196 (0.24623)	−0.46352*** (0.14641)	0.28658 (0.29576)
Democratic governor	0.65509** (0.30788)	−0.00992 (0.25438)	−0.92443** (0.37021)
Percent state Democratic legislators	0.01203 (0.02448)	−0.02851 (0.03756)	−0.04811 (0.03523)
R^2 within between overall	0.1318 0.0559 0.0554	0.0954 0.0163 0.0091	0.0551 0.0379 0.0380
Observations	588	588	588

Note FE means fixed-effects model and IV means instrumental variable regression. Cluster-robust standard errors are in parentheses. Estimation uses data from 49 states between 2005 and 2016. ***Significant at 1% level, **Significant at 5% level, *Significant at 10% level

to states with good weather.[4] The data on annual average temperature by state come from the National Centers for Environmental Information.[5]

Table 6.2 reports the estimates. A fixed-effects model is adopted.[6] We find that a higher share of the elderly is associated with a lower share of the corporate income tax and sales tax, though it is not statistically significant for the sales tax. These results are consistent with our prediction, but the result on the sales tax is weaker than expected.

---

[4] See Sect. 5.3.1 on the relation of average temperature and elderly population ratio.

[5] As explained in Sect. 5.3.1, the statewide data for Hawaii are unavailable, so that we use city-level annual average temperature of Honolulu International Airport.

[6] We conducted panel data analysis without instrumenting the percent elderly population, to select fixed-effects versus random-effects models. The result of the test using the Sargan-Hansen statistic requires us to select fixed-effects model.

6.4 Minimum Wage

We examine state policies on the minimum wage, using reduced-form reaction functions. A reaction function relates the choice of a state government to the choices of the governments in neighboring states, controlling various characteristics of the state. The slope of the reaction function indicates the extent to which state governments compete with each other. Besides strategic interdependence, we are interested in the effect of a state's elderly population.

We first estimate a spatial static panel data model. The reaction function for a state government competing with other governments is formalized as

$$mw_{it} = \alpha_i + \delta \sum_{j \neq i} \omega_{ij} mw_{jt} + x'_{i,t-1}\beta + z'_{i,t-1}\gamma + \varepsilon_{it}, \tag{6.2}$$

where mw_{it} is the minimum wage in state i in year t; α_i is a time-invariant state-specific effect; δ is the slope parameter; ω_{ij} is the exogenously determined weight, normalized so that $\sum_{j \neq i} \omega_{ij} = 1$; $x_{i,t-1}$ denotes a vector of explanatory variables representing demographic characteristics of state i in year $t-1$; $z_{i,t-1}$ denotes a vector of other control variables describing economic and political factors; β and γ are vectors of parameters. In regard to $x_{i,t-1}$ and $z_{i,t-1}$, we take a one-year lag to enhance exogeneity, which is a stylized manner in the literature. An error term, ε_{it}, completes the equation.

In (6.2), the minimum wages set by competitors are represented by the spatial lag term, $\sum_{j \neq i} \omega_{ij} mw_{jt}$, which is the weighted average of minimum wages in other states. We use three different weighting schemes. In the literature, geographic criteria are frequently adopted. One of the reasons is that they yield purely exogenous weights.

The first is constructed on the basis of contiguity of states, that is, whether they share a common border:

$$\omega_{ij} = \frac{b_{ij}}{\sum_{h \neq i} b_{ih}}, \tag{6.3}$$

in which b_{ij} is a border dummy which takes 1 when states i and j share a common border, and 0 otherwise.

The second is an inverse-distance weight, which allows closely located states to have stronger effects on each other than states located farther away. The inverse-distance weight is given by

$$\omega_{ij} = \frac{\frac{1}{d_{ij}}}{\sum_{h \neq i} \frac{1}{d_{ih}}}, \tag{6.4}$$

in which d_{ij} is the distance between the most populated cities in states i and j.

The inverse-distance weight does not capture economic closeness. Among states that are equally distant, businesses may be more attracted by the state having many opportunities of employment and consumption. The third weight reflects this element by multiplying $\frac{1}{d_{ij}}$ or $\frac{1}{d_{ih}}$ in (6.4) by the total population in state j or h. We assume in calculation of this weight that the population in each state had been fixed at the level in 2010.

In Eq. (6.2) the minimum wage in the state and the minimum wages of competitors are assumed to be simultaneously determined. This means that the weighted average of competitors' minimum wages is endogenous. To address this endogeneity, we use an instrumental variable, which yields consistent estimates even in the case of spatial error dependence. More specifically, the weighted minimum wages are instrumented by the weighted average of percent Democratic legislators in other states (lagged one year to avoid estimation problems arising from simultaneity). The competing states' political preferences will only affect their own minimum wages through politics, but not minimum wages in other states. Political preferences of state residents, while unobserved, should be largely revealed by electoral outcomes.[7]

We incorporate dynamics by including a lagged dependent variable in the right-hand side of the reaction function (6.2) and employing the Arellano and Bond (1991) Dynamic Panel Data estimator. The Arellano and Bond (1991) Dynamic Panel Data estimator is a Generalized Method of Moments (GMM) estimator, which addresses endogeneity. Instruments are differenced terms on the right-hand side, all at their lagged levels.

Formally, by adding the time-lagged minimum wage in state i on the right-hand side of (6.2), the model becomes dynamic:

$$mw_{it} = \alpha_i + \mu mw_{i,t-1} + \delta \sum_{j \neq i} \omega_{ij} mw_{jt} + x'_{i,t-1}\beta + z'_{i,t-1}\gamma + \varepsilon_{it}. \tag{6.5}$$

Written in first differences, unobserved state-specific effect, α_i, is eliminated:

$$\Delta mw_{it} = \mu \Delta mw_{i,t-1} + \delta \Delta \sum\nolimits_{j \neq i} \omega_{ij} mw_{jt} + \Delta x'_{i,t-1}\beta + \Delta z'_{i,t-1}\gamma + \Delta\varepsilon_{it}, \tag{6.6}$$

in which $\Delta mw_{it} \equiv mw_{it} - mw_{i,t-1}$ is the change in the minimum wage in state i between years $t - 1$ and t, and the same usage applies to other variables.

The panel of state-level data covers 48 states from 2005 to 2016. Alaska and Hawaii are excluded from our samples because these two states do not share a border with any US state and are geographically distant to other states in the US. The data on minimum wages come from Sorens et al. (2008) and the updated database.[8] Note that the minimum wage in a state is calculated as the sum of the minimum wage set by the federal government and the additional part set by the state government. Therefore,

[7] We follow Chirinko and Wilson (2017) who use the governorship as an instrument.

[8] The updated data are available at http://www.statepolicyindex.com.

if the state government adopts no additional legislation, the minimum wage in that state is equal to the minimum wage set by the federal government.

Demographic variables included in $x_{i,t-1}$ are percent population aged 65 and over and the total population (in one hundred thousands). The variables included in $z_{i,t-1}$ are classified into economic and political variables. Economic variables are unemployment rate, real GDP per capita, and real personal income per capita. Following Egger et al. (2005) and Devereux et al. (2007), we use percent Democratic legislators which represent a state's political inclination, as a political variable. Democratic legislators are expected to favor a higher minimum wage. The data sources of these variables are the same as in Sect. 6.3.

Table 6.3 reports the results from the static model. Whichever weight is used among the three measures of distance, we find that a state's minimum wage increases with the minimum wages of neighboring states. Regarding the percent elderly, the estimated coefficient is insignificant, and its sign varies with the weight adopted. The fraction of the population in a state that is elderly has only a weak effect on the minimum wage.

Table 6.3 Effects of the elderly population on minimum wage (static model)

Weighting matrix	Contiguity(RE-IV)	Inverse-distance (FE-IV)	Population/distance (RE-IV)
Neighbors' minimum wage	0.66907** (0.25621)	1.05177*** (0.20171)	1.05626*** (0.28390)
Percent aged 65 and over 1-year lag	0.06672 (0.04223)	−0.01468 (0.04874)	0.02170 (0.04955)
Population 1-year lag	0.00164** (0.00070)	0.00046 (0.00481)	0.00148 (0.00098)
Percent unemployed 1-year lag	−0.00453 (0.01423)	−0.01032 (0.01566)	0.02292 (0.01520)
Real GDP per capita 1-year lag	0.00249 (0.00930)	0.00865 (0.00947)	0.00816 (0.00741)
Real personal income per capita 1-year lag	0.01616 (0.02601)	−0.02341 (0.03987)	0.02413 (0.02307)
Percent state Democratic legislators 1-year lag	0.00485 (0.00300)	0.00053 (0.00350)	0.00545** (0.00245)
R^2 within between overall	0.4668 0.4926 0.4820	0.4655 0.2231 0.3003	0.4316 0.5197 0.4815
Observations	528	528	528

Note FE represents fixed-effects model and RE represents random-effects model. IV means instrumental variable regression. Cluster-robust standard errors are in parentheses. Estimation uses data from 48 states between 2005 and 2016. ***Significant at 1% level, **Significant at 5% level, *Significant at 10% level

Table 6.4 Effects of the elderly population on minimum wage (dynamic model)

Weighting matrix	Contiguity	Inverse-distance	Population/distance
Neighbors' minimum wage	0.44203*** (0.05767)	0.55654*** (0.07821)	0.59154*** (0.08559)
Minimum wage 1-year lag	0.49272*** (0.05502)	0.51181*** (0.05337)	0.52455*** (0.05367)
Percent aged 65 and over 1-year lag	0.02149 (0.01759)	−0.00273 (0.01896)	−0.01466 (0.02004)
Population 1-year lag	0.00860 (0.00626)	0.00740 (0.00628)	0.01015 (0.00637)
Percent unemployed 1-year lag	−0.07868*** (0.01060)	−0.08197*** (0.01053)	−0.06739*** (0.01109)
Real GDP per capita 1-year lag	−0.01363 (0.00840)	−0.00545 (0.00862)	−0.00633 (0.00869)
Real personal income per capita 1-year lag	−0.06662*** (0.02544)	−0.09426*** (0.02682)	−0.07342*** (0.02627)
Percent state Democratic legislators1-year lag	0.01050*** (0.00214)	0.01021*** (0.00219)	0.01038*** (0.00222)
Wald test	Chi2(8) = 788.32 Prob > chi2 = 0.0000	Chi2(8) = 761.69 Prob > chi2 = 0.0000	Chi2(8) = 732.22 Prob > chi2 = 0.0000
Observations	432	432	432

Note Standard errors are in parentheses. Estimation uses data from 48 states between 2005 and 2016. ***Significant at 1% level, **Significant at 5% level, *Significant at 10% level

Table 6.4 reports the results from the dynamic model. It also shows evidence for interaction among state governments. The estimated coefficient is approximately 0.5 for all weights. Thus, the estimated reaction coefficient by applying Dynamic Panel Data estimation is smaller than the estimate by the static-model regression. This is consistent with the result in Jacobs et al. (2010). We also find inertia; the minimum wage in the previous year affects the current minimum wage, suggesting that minimum wage is persistent over time.

We find no statistically significant effect of the share of the elderly on the minimum wage. This result is different from the result in Sect. 6.3 that a higher share of the elderly population is associated with a smaller share of corporate income tax revenue.

Among other control variables, higher unemployment rate leads to a lower minimum wage. A higher share of Democratic legislators is associated with a higher minimum wage, indicating their disinclination to fluctuation in labor income.

6.5 Conclusion

We examined how the share of the elderly in a state, which is correlated with the elderly's political power, affects a state government's policy on taxation and intervention in the labor market.

Our estimation looked at the share of tax revenues from various sources as dependent variables, using weather as an instrument for the elderly proportion in a state. We find that as the elderly constitute a larger share of the population, the share of a state's revenue coming from the corporate income tax declines. This suggests that elderly citizens are concerned about indirect effects from increased tax burden on firms, on employment and consumer prices. Furthermore, the elderly own more stocks than others, and so care more about corporate profits. For example, in 2016 householders aged 65 and over had a median value of stock ownership of $100,000, compared to only $70,000 for those aged 55–64, and only $50,000 for those aged 45–54.[9]

Though the size of the elderly population affected corporate income taxes in a state, our estimates find no significant effect of the elderly population on the minimum wage. How can we reconcile these results? The elderly may not be farsighted, but our estimation results on taxation are consistent with concern about increased corporate income taxes, which can affect their consumption. Indeed, our regression using the share of sales tax revenue as a dependent variable gives us a weak but negative effect of the share of elderly population. Interestingly, their concerns on the corporate income tax are more apparent. Our explanation in Sect. 6.1 may be applicable to these different effects. If employers have monopsony power in the local labor market for low-wage workers, the minimum wage may make higher wages little affect the prices of goods and services. The elderly then may not strongly oppose the local government's higher minimum wage setting.

References

Aaronson D, French E, MacDonald J (2008) The minimum wage, restaurant prices, and labor market structure. J Hum Resour 43(3):688–720

Arellano M, Bond S (1991) Some tests of specification for panel data: Monte Carlo evidence and an application to employment equations. Rev Econ Stat 58(2):277–297

Azar J, Huet-Vaughn E, Marinescu I, Taska B, von Wachter T (2019) Minimum wage employment effects and labor market concentration. NBER Working Paper 26101

Breuckner J (2003) Strategic interaction among governments: An overview of theoretical studies. Int Reg Sci Rev 26(2):175–188

Cadena BC (2014) Recent immigrants as labor market arbitrageurs: evidence from the minimum wage. J Urban Econ 80:1–12

Card DE, Krueger AB (1995) Myth and measurement: the new economics of the minimum wage. Princeton University Press, Princeton

---

[9] The data are collected by Survey of Income and Program Participation, US Census Bureau, and retrieved from: https://www.census.gov/data/tables/2016/demo/wealth/wealth-asset-ownership.html.

Chirinko RS, Wilson DJ (2017) Tax competition among U.S. states: racing to the bottom or riding on a seesaw? J Public Econ 155:147–163
Davies RB, Vadlamannati KC (2013) A race to the bottom in labor standards? An empirical investigation. J Dev Econ 103:1–14
Devereux M, Lockwood B, Redoano M (2007) Horizontal and vertical indirect tax competition: theory and some evidence form the USA. J Public Econ 91(3–4):451–479
Egger p, Pfaffermayr M, Winner H, (2005) Commodity taxation in a 'linear' world: a spatial panel data approach. Reg Sci Urban Econ 35(5):527–541
Hirsch BT, Schumacher EJ (1995) Monopsony power and relative wages in the labor market for nurses. J Health Econ 14(4):443–476
Jacobs JPAM, Ligthart JE, Vrijburg H (2010) Consumption tax competition among governments: evidence from the United States. Int Tax Public Finance 17(3):271–294
Li YL, Kanbur RK, Lin CL (2019) Minimum wage competition between local governments in China. J Dev Stud 55(12):2479–2494
Matsudaira JD (2014) Monopsony in the low-wage labor market? Evidence from minimum nurse staffing regulations. Rev Econ Stat 96(1):92–102
Monras J (2019) Minimum wages and spatial equilibrium: theory and evidence. J Labor Econ 37(3):853–904
Sorens J, Muedini F, Ruger WP (2008) State and local public policies in 2006: a new database. State Polit Policy Q 8(3):309–326
Staiger DO, Spetz J, Phibbs CS (2010) Is there monopsony in the labor market? Evidence from a natural experiment. J Labor Econ 28(2):211–236
Stigler GJ (1946) The economics of minimum wage legislation. Am Econ Rev 36(3):358–365

Chapter 7
Conclusion

7.1 Generational Conflict

This book has explored generational conflict, in the context of governmental spending. The problem is more general, as the Covid-19 pandemic has highlighted. Should vaccines at first be given only to the elderly, or should eligibility be extended to essential workers and to school teachers, who are usually younger. Lockdowns protect the elderly, but hurt the young. Same with school re-openings. An intergenerational investment problem arises with policies to address global warming: spending today imposes taxes on the elderly, but would bring benefits largely in the future, when the current elderly are not alive.

Generational conflict has been studied by others. Danish data indicate that an increase in the elderly population reduces spending on child care and education, but that an increase in the number of younger voters does not reduce spending on elderly care (Borge and Rattsø 2008). In Switzerland, elderly voters are less willing to support spending on education, preferring spending on health and social security (Cattaneo and Wolter 2009). Panel data for the states of the US over the 1960–1990 period finds that an increase in the fraction of elderly residents in a jurisdiction is associated with reduced per-child educational spending (Poterba 1997).

With these issues, and with public spending, the common assumption is that people are selfish. In most of this book we had assumed, explicitly or implicitly, that a person favored the policy that, in a direct manner, most benefits him. The elderly want government services that are, at least partly, paid by the young. The young oppose spending on the elderly. We did at times consider strategic behavior, such as the elderly favoring services valued by the young, to encourage the young to live in locales inhabited by the elderly, and so reducing wages paid to the young. But that too assumes selfishness.

K. Terai et al., *The Political Economy of Population Aging*, Advances in Japanese Business and Economics 30, https://doi.org/10.1007/978-981-16-5536-4_7

7.2 Altruism

A question is whether political choices would differ if people, both the young and the old, are altruistic. The elderly are often parents of the young, and children may also care about the well-being of their parents. Some evidence for altruism is given by Logan and Spitze (1995): a survey of individuals shows that older people are the least likely to support governmental programs for older people. How would such altruism cause public policy to respond to an increase in the number of the elderly in a locale?

7.2.1 Parents Care About Their Children

If parents care about their children, then we would expect that an increase in the number of the elderly in a locale would cause little decline in per capita spending on education, or cause little increase in spending on the elderly. But the size of that effect can depend on several factors. Because of decreasing marginal utility of income, a parent who has many children, about whom he cares, will be more willing to tax them to support himself than if he has few children: $100 more spent on the parent will be spread among few of his or her children, and so cause a large loss in utility.

It also matters if the children live in the same locale as the parents. We can imagine one region, A, with 1000 elderly and 1000 middle-aged people, but none of the middle aged are children of the elderly. In another region, B, the age distribution is the same, but all of the middle-aged people are children of the elderly. Then with altruism by parents, we would expect less spending on education, and more spending on the elderly, in region A than in region B. We would also expect that mobility, of either the elderly or their children, will affect these different patterns. High mobility will lead to the pattern in A; low mobility to the pattern in B. So, to the extent that the US has high mobility by both the elderly and their children, or to the extent that in Japan the elderly stay put while their children move to large cities, the elderly will appear selfish.

Altruism by the elderly will affect what policies that benefit the young the elderly prefer. There should be greater support for capital investments that serve the elderly: they benefit the current elderly and the future elderly. So, we may see excessive investment in hospitals. Another capital investment is the development of pharmaceuticals, which will also help the people who will later be elderly.

7.2.2 Bargaining

Interestingly, altruism reflected in paternalism can cause what looks like selfish behavior. A mother may want her daughter to avoid drug use, marry within the religion, or complete a college degree. The mother can offer cash or the promise of a bequest in return for the behavior of the child the mother wants to incentivize. Thus, we follow Bernheim et al. (1985), Cremer and Pestieau (1991), Cox and Rank (1992), and Cremer et al. (1992) in supposing that parents leave bequests to their children because they want to influence their children.

Bernheim, Shleifer, and Summers (1985) provide strong evidence for the strategic bequest motive, finding in particular that attention to parents by children increases with bequeathable wealth, but decreases with non-bequeathable wealth. Also consistent with the strategic bequest model, Borsch-Supan et al. (1992) find that children who earn higher wages spend less time with their elderly parents. Laitner and Ohlsson (2001) study Sweden and the US, finding that parental bequests increase with the parents' lifetime resources, and decline with the earnings potential of the heir. These results are consistent with both an altruistic motive and a strategic bequest motive, but, the authors state, perhaps better fit the strategic bequest motive. Hochguertel and Ohlsson (2000), who examine pre-death gifts made by parents, also reach conclusions consistent with the strategic bequest model: in the US a child is more likely to receive a gift if she works fewer hours and has lower income than her brothers and sisters.

Some works, however, question the hypothesis: in the US children's provision of care to parents is little guided by a strategic bequest motive. See Perozek (1998) who, unlike Bernheim et al. (1985), controls for the number of children in a family, and Sloan et al. (1997) who study the amount of time children devote to disabled elderly parents.

The strategic bequest motive appears far stronger outside the US. Horioka et al. (2000) compare the responses of survey respondents in the US and Japan. In one question, respondents were asked whether no strategic considerations entered in bequests. In the US 43 percent of respondents held the view "I want to make efforts to leave behind a bequest regardless of whether my child or children look after me after I retire;" in Japan only 20 percent of respondents agreed with this non-selfish view. In Japan 33 percent of respondents said that "Most or all of [the bequest] will be willed to the child or children who look after me;" only two percent held this view in the US.

Consider a public project which produces a consumption good and which benefits future generations. Let a conventional cost–benefit analysis find that it gives higher benefits than projects it would displace in the private sector. Elderly voters may nevertheless oppose the public project: the combination of a desire to control bequests and the lack of control over who gets benefits from a public project makes the elderly disfavor the public project. In contrast, private projects have owners, allowing parents to control whether their children will receive the benefits from such projects. Parents can therefore better influence the behavior of their children when they have the option of giving the children title to private assets. One example is housing: the elderly may prefer privately owned housing, even if expensive, which they can bequeath, over publicly provided housing. Or the elderly may not want publicly provided child care, instead wanting their children to depend on grandparents to provide care and to give cash to the grandchildren.

Or consider a parent evaluating a tax increase to fund a government program that would reduce carbon emissions. The program has no immediate effects, but will benefit the next generation. Suppose that standard cost-benefit analysis, which compares the present discounted value of social costs to benefits, calls for undertaking the program. Let consecutive generations, say parent and child, implicitly bargain over how much service the child provides the parent in exchange for a bequest. A parent may oppose even a costless public investment that benefits the child. The opposition arises because the benefits of the public project increase the child's income, which increases the child's utility under the threat point when bargaining with the parent, and so reduces the parent's ability to influence the behavior of her children.

7.2.3 *Children Care About Their Parents*

Consider next altruism by children to their elderly parents. We must note that what may look like altruism can instead be motivated by selfish interests. For example, middle-aged children may want governmental assistance for home care, so that the children don't have to provide it. Or, taxes which fund services to the elderly may reduce spending by the elderly, and so increase bequests that the parents leave.

7.3 Expressive Voting

Expressive voting may also lead to such a pattern. Children of the elderly may be ashamed to tell their parents that they voted against spending on the elderly, and so vote for it. Because no one vote is decisive, a middle-aged voter may vote against his own self-interest. In geographically small countries, people see their parents and grandparents more often, and so the effect will be greater there. Other conditions can also lead to different results. For example, an adult child may have greater desire to tell her parents that she voted for elderly care if she frequently sees them in person, than if she rarely does. So residential mobility can affect voting.

The evidence on expressive compared to instrumentalist voting is mixed. Tyran (2004) tests a model of expressive voting by experimentally investigating a proposal to tax everyone and to donate tax revenues, finding little support for the theory, instead finding that voters tend to approve the proposal if they expect others to approve, too. Experimental subjects who could vote on having some cash given to themselves or instead to charity were more likely to vote for the charity the less likely their votes were decisive; the effect is weak, but consistent with expressive voting (Carter and Guerette 1992). Lastly, instrumentalist theory claims that a person is more likely to vote the more strongly he believes that his vote will make a difference. This hypothesis is not supported by the available data (see Ashenfelter and Kelley 1975).

7.4 Extensions to Other Countries

What can we learn about countries other than Japan and the US? Europe and China are also aging. One lesson is that the effects differ across countries. We had discussed how residential mobility can affect altruism. In addition, aging is not uniform across different groups. For example, in the US, Blacks and Hispanics are younger than are Whites, so spending on the elderly disproportionately benefits Whites. That in turn affects political support for different policies. Similarly, women constitute a larger fraction of the elderly than do men, so we would expect men to show lower support for services to the elderly. On the other hand, evidence shows that mothers care more their children than do men, and so in countries with a larger fraction of the elderly who are female, support for services that benefit the young may be larger.

References

Ashenfelter O, Kelley S Jr (1975) Determinants of participation in presidential elections. J Law Econ 18(3):695–733

Bernheim BD, Shleifer A, Summers LH (1985) The strategic bequest motive. J Political Econ 93(6):1045–1076

Borge LE, Rattsø J (2008) Young and old competing for public welfare services. CESifo Working Paper 2223

Borsch-Supan A, Gokhale J, Kotlikoff LJ, Morris JN (1992) The provision of time to the elderly by their children. In: Wise DA (ed) Topics of the economics of aging. University of Chicago Press, Chicago

Carter JR, Guerette SD (1992) An experimental study of expressive voting. Public Choice 73(3):251–260

Cattaneo MA, Wolter SC (2009) Are the elderly a threat to educational expenditures? Eur J Political Econ 25(2):225–236

Cox D, Rank MR (1992) Inter-vivos transfers and intergenerational exchange. Rev Econ Stat 74(2):305–314

Cremer H, Pestieau P (1991) Bequests, filial attention and fertility. Economica 58(231):359–375

Cremer H, Kessler D, Pestieau P (1992) Intergenerational transfers within the family. Eur Econ Rev 36(1):1–16

Hochguertel S, Ohlsson H (2000) Compensatory inter vivos gifts. Working paper in economics, Department of Economics, Goteborg University

Horioka, CY, Fujisaki H, Watanabe W, Kouno T (2000) Are Americans more altruistic than the Japanese? A U.S.-Japan comparison of saving and bequest motives. Int Econ J 14(1):1–31

Laitner J, Ohlsson H (2001) Bequest motives: a comparison of Sweden and the United States. J Public Econ 79(1):205–236

Logan JR, Spitze GD (1995) Self-interest and altruism in intergenerational relations. Demography 32(3):353–364

Perozek MG (1998) Comment: a reexamination of the strategic bequest motive. J Political Econ 106(2):423–445

Poterba JM (1997) Demographic structure and the political economy of public education. J Policy Anal Manag 16(1):48–66

Sloan FA, Picone G, Hoerger TJ (1997) The supply of children's time to disabled elderly parents. Econ Inq 35(2):295–308

Tyran J-R (2004) Voting when money and morals conflict: an experimental test of expressive voting. J Public Econ 88(7–8):1645–1664

Index

K. Terai et al., *The Political Economy of Population Aging*, Advances in Japanese Business and Economics 30, https://doi.org/10.1007/978-981-16-5536-4